Statistical Physics, Automata Networks and Dynamical Systems

Mathematics and Its Applications

Volume 75

Statistical Physics, Automata Networks and Dynamical Systems

edited by

Eric Goles

and

Servet Martínez

Department of Mathematical Engineering,
University of Chile,
Santiago, Chile

KLUWER ACADEMIC PUBLISHERS

DORDRECHT / BOSTON / LONDON

Library of Congress Cataloging-in-Publication Data

Statistical physics, automata networks, and dynamical systems / edited by Eric Goles and Servet Martínez.
p. cm. -- (Mathematics and its applications ; v. 75)
Grew out of courses given at the Second School on Statistical Physics and Cooperative Systems held Dec. 10-14, 1990, in Santiago, Chile.
ISBN 0-7923-1595-2 (HB : acid free paper)
1. Statistical physics. 2. Cellular automata. 3. Neural networks (Computer science) I. Golès, E. II. Martínez, Servet. III. School on Statistical Physics and Cooperative Systems (2nd : 1990 : Santiago, Chile) IV. Series: Mathematics and its applications (Kluwer Academic Publishers) ; v. 75.
QC174.8.S73 1992
530.1'3--dc20 92-3913

ISBN 0-7923-1595-2

Published by Kluwer Academic Publishers,
P.O. Box 17, 3300 AA Dordrecht, The Netherlands.

Kluwer Academic Publishers incorporates
the publishing programmes of
D. Reidel, Martinus Nijhoff, Dr W. Junk and MTP Press.

Sold and distributed in the U.S.A. and Canada
by Kluwer Academic Publishers,
101 Philip Drive, Norwell, MA 02061, U.S.A.

In all other countries, sold and distributed
by Kluwer Academic Publishers Group,
P.O. Box 322, 3300 AH Dordrecht, The Netherlands.

Printed on acid-free paper

Printed in the Netherlands

SERIES EDITOR'S PREFACE

'Et moi, ..., si j'avait su comment en revenir, je n'y serais point allé.'
Jules Verne

The series is divergent; therefore we may be able to do something with it.
O. Heaviside

One service mathematics has rendered the human race. It has put common sense back where it belongs, on the topmost shelf next to the dusty canister labelled 'discarded nonsense'.
Eric T. Bell

Mathematics is a tool for thought. A highly necessary tool in a world where both feedback and nonlinearities abound. Similarly, all kinds of parts of mathematics serve as tools for other parts and for other sciences.

Applying a simple rewriting rule to the quote on the right above one finds such statements as: 'One service topology has rendered mathematical physics ...'; 'One service logic has rendered computer science ...'; 'One service category theory has rendered mathematics ...'. All arguably true. And all statements obtainable this way form part of the raison d'être of this series.

This series, *Mathematics and Its Applications*, started in 1977. Now that over one hundred volumes have appeared it seems opportune to reexamine its scope. At the time I wrote

> "Growing specialization and diversification have brought a host of monographs and textbooks on increasingly specialized topics. However, the 'tree' of knowledge of mathematics and related fields does not grow only by putting forth new branches. It also happens, quite often in fact, that branches which were thought to be completely disparate are suddenly seen to be related. Further, the kind and level of sophistication of mathematics applied in various sciences has changed drastically in recent years: measure theory is used (non-trivially) in regional and theoretical economics; algebraic geometry interacts with physics; the Minkowsky lemma, coding theory and the structure of water meet one another in packing and covering theory; quantum fields, crystal defects and mathematical programming profit from homotopy theory; Lie algebras are relevant to filtering; and prediction and electrical engineering can use Stein spaces. And in addition to this there are such new emerging subdisciplines as 'experimental mathematics', 'CFD', 'completely integrable systems', 'chaos, synergetics and large-scale order', which are almost impossible to fit into the existing classification schemes. They draw upon widely different sections of mathematics."

By and large, all this still applies today. It is still true that at first sight mathematics seems rather fragmented and that to find, see, and exploit the deeper underlying interrelations more effort is needed and so are books that can help mathematicians and scientists do so. Accordingly MIA will continue to try to make such books available.

If anything, the description I gave in 1977 is now an understatement. To the examples of interaction areas one should add string theory where Riemann surfaces, algebraic geometry, modular functions, knots, quantum field theory, Kac-Moody algebras, monstrous moonshine (and more) all come together. And to the examples of things which can be usefully applied let me add the topic 'finite geometry'; a combination of words which sounds like it might not even exist, let alone be applicable. And yet it is being applied: to statistics via designs, to radar/sonar detection arrays (via finite projective planes), and to bus connections of VLSI chips (via difference sets). There seems to be no part of (so-called pure) mathematics that is not in immediate danger of being applied. And, accordingly, the applied mathematician needs to be aware of much more. Besides analysis and numerics, the traditional workhorses, he may need all kinds of combinatorics, algebra, probability, and so on.

In addition, the applied scientist needs to cope increasingly with the nonlinear world and the extra mathematical sophistication that this requires. For that is where the rewards are. Linear models are honest and a bit sad and depressing: proportional efforts and results. It is in the nonlinear world that infinitesimal inputs may result in macroscopic outputs (or vice versa). To appreciate what I am hinting at: if electronics were linear we would have no fun with transistors and computers; we would have no TV; in fact you would not be reading these lines.

There is also no safety in ignoring such outlandish things as nonstandard analysis, superspace and anticommuting integration, *p*-adic and ultrametric space. All three have applications in both electrical engineering and physics. Once, complex numbers were equally outlandish, but they frequently proved the shortest path between 'real' results. Similarly, the first two topics named have already provided a number of 'wormhole' paths. There is no telling where all this is leading - fortunately.

Thus the original scope of the series, which for various (sound) reasons now comprises five subseries: white (Japan), yellow (China), red (USSR), blue (Eastern Europe), and green (everything else), still applies. It has been enlarged a bit to include books treating of the tools from one subdiscipline which are used in others. Thus the series still aims at books dealing with:

- a central concept which plays an important role in several different mathematical and/or scientific specialization areas;
- new applications of the results and ideas from one area of scientific endeavour into another;
- influences which the results, problems and concepts of one field of enquiry have, and have had, on the development of another.

Let us list some of the more pervasive buzz-words of the moment in mathematics and its neighboring sciences: fractal structures, neural nets and nets of automata, turbulence, spin glasses, statistical learning theory, strange attractors, topological entropy, Kolmogorov scaling, shock waves, laws of large numbers, exclusion processes, chaos. All are very much part of modern nonlinear mathematics. Are they all linked and related? Yes, in many ways; too many to cover in one volume.

But in this volume edited by Golez and Martinez, authors of one of the early influential monographs on neural nets (Neural and automata networks, KAP, 1990), quite a few of these interconnections are treated. Through the medium of six surveys by leading scientists, the reader has the opportunity to learn a great deal about these topics where physics, mathematics and computer science interact.

The shortest path between two truths in the real domain passes through the complex domain.

J. Hadamard

La physique ne nous donne pas seulement l'occasion de résoudre des problèmes ... elle nous fait pressentir la solution.

H. Poincaré

Never lend books, for no one ever returns them; the only books I have in my library are books that other folk have lent me.

Anatole France

The function of an expert is not to be more right than other people, but to be wrong for more sophisticated reasons.

David Butler

Bussum, 9 February 1992

Michiel Hazewinkel

FOREWORD

This book grew out of courses given at the Second School on Statistical Physics and Cooperative Systems held at Santiago, Chile, 10th to 14th December 1990. The main idea of this periodic school was to bring together scientists working on subjects related with recent trends in Statistical Physics. More precisely in subjects related with non linear phenomena, dynamical systems, fractal structures, cellular automata, exclusion processes, and neural networks. Scientists working in these subjects come from several areas: computer science and artificial intelligence, mathematics, biology, and physics. Recently, a very important cross-fertilization has taken place with regard to the aforesaid scientific and technological disciplines, so as to give a new approach to the research whose common core remains in statistical physics.

Each contribution is devoted to one or more of the previous subjects. In most cases they are structured as surveys, presenting at the same time an original point of view about the topic and showing mostly new results.

The expository text of Pierre Collet is devoted to dynamical systems. The different kinds of behavior appearing in physical systems are discussed in it, as well as the main ideas on stability, elementary bifurcations and the period doubling road to turbulence. Chaotical systems are reviewed in relation with the Sinai-Ruelle-Bowen measure, the dimension of attractors and topological entropy. A final discussion on experimental measurement of the quantities involved in chaotical systems is made.

The survey paper of Pablo Ferrari concerns the Burgers equation, which constitutes a model of transport on highways. This equation is analysed from a microscopical point of view, more precisely it is studied as a limit of totally symmetric simple exclusion processes. The usual ideas of particle systems are introduced, in particular coupling techniques. The main results contain the description of the set of invariant measures of the exclusion process, the existence of shocks at microscopical level and laws of large numbers and central limit theorems for the shock. The last part of the paper discusses some other probabilistic cellular automata.

Two examples of fractals are presented by Hans Herrmann. The former is a deterministic model for a dense packing of disks rolling on each other which might there applications in turbulence and tectonic motion. The fractal dimension is calculated and compared with the Kolmogorov scaling. The second example is related with the dynamics of a probabilistic cellular automation associated with local changes in Ising configurations. The damage between two configurations differing in only one site is submitted to the Monte-Carlo dynamics. It is shown that for a critical value of the temperature parameter there appear damage avalanches obeying a power-law size distribution and that the fractal dimension of the damage cluster can be related exactly to critical exponents for the heath bath dynamics.

The survey works of Pierre Peretto, Mirta Gordon and Miguel Rodríguez-Girones study neural networks in the framework of statistical physics. In particular, the dynamics of neural networks are derived from statistical mechanics considerations. This approach also serves to introduce learning theory, and to discuss the Boltzmann networks like neocognition and back-propagation algorithm.

In the paper of Eric Goles and Servet Martínez, the relation between automata networks and optimization problems is studied. Several problems on optimization can be seen as maximization of subnorms restricted to a suitable unit ball of a linear space. It is shown that for this kind of problems the "positive" ($\underline{P}$) automata dynamics constitutes a deterministic hill-climbing strategy converging to local optima. This dynamics contains the classical iterated power algorithm to compute spectral norms. In this approach, restrictions form part of the dynamics which differs from the Hopfield-Tank approach where the restrictions appear as a quadratic perturbation of the cost function.

The review paper of Pierre Picco concerns solvable models associated to spin glasses. The REM and GREM constitute a simplification of the Sherrington-Kirkpatrick model. The energy variable is also Gaussian in them but the covariance matrix is simpler: it is either the identity or a hierarchized matrix. In this work, the free energy of the GREM is fully described, as well as the relation between the number of categories and the volume. A detailed proof of these results is presented.

The editors are grateful to the participants of the School, as well as to the authors of the individual chapters. They are also indebted to the sponsors: Fundación Andes, Fondecyt, Conicyt, Cooperación Francesa, Departamento de Relaciones Internacionales and DTI of the Universidad de Chile, Centro Internacional de Física (Colombia) and the Departamento de Ingeniería Matemática and Cenet of the Facultad de Ciencias Físicas y Matemáticas.

Mrs. Gladys Cavellone deserves a special mention for her very fine and hard work typing the book.

The Editors

REGULAR AND CHAOTIC BEHAVIOUR OF DYNAMICAL SYSTEMS

P. COLLET
Centre de Physique Théorique
Laboratoire CNRS UPR 14
Ecole Polytechnique
F-91128 Palaiseau Cedex
France

1. Introduction

The theory of dynamical systems is devoted to the study of time evolutions. The instantaneous state of a physical system is described by a point in some phase space which is usually a vector space or a manifold. The time evolution is then given by a vector field (continuous time) or a mapping (discrete time). The main goal is to understand qualitatively and if possible quantitatively the long time behaviour of these systems.

This is a very general and important problem which is met in various areas of Physics: classical mechanics, hydrodynamics, meteorology..., and in many other sciences as well. In all these examples the time evolution is described by a differential equation in time (possibly a partial differential equation). We will always assume that this equation can be regarded as given by a vector field on some phase space. For example equations which are higher order in time can be reduced to coupled systems of first order equations. There may be also some constraints which impose a non trivial topology to the phase space. We also want to include in our study the case of infinite dimensional vector fields which appear for example in the case of the Navier Stokes equation in a finite domain, where the phase space is the space of all divergenceless vector fields which satisfy the boundary conditions.

Discrete time iterations appear naturally in a number of situations. The simplest one is when one is using a stroboscope, which amounts to the observation of the system at periodic time instants. This technique is of course well adapted to the study of periodic time evolutions and the rigorous version is the Poincaré section [R]. A related situation is the case of vector fields which depend periodically on time. By using an auxiliary variable, one can transform non autonomous systems into autonomous ones in a phase space with one more dimension. In the case of periodic time dependence, it is often better to consider the evolution of the system over one period and to discuss the dynamics of this mapping.

E. Goles and S. Martínez (eds.), Statistical Physics, Automata Networks and Dynamical Systems, 1–23.

In order to introduce some of the fundamental tools in the study of dynamical systems we will first consider the well known (and rather trivial) example of the harmonic oscillator. We first observe that the phase space is the two dimensional plane, and the time evolution is given by the Hamiltonian vector field, i.e. a coupled system of two differential equations. The trajectories in phase space are ellipses of constant energy (given by the initial condition) and the motion is a rotation around the ellipse. In that case we have solved the problem of finding the large time dynamical behaviour of our system: it is just this trivial motion. Let us now add some damping proportional to the momentum. The motion is still described by a vector field in $\mathbf{R}^2$ but now the system is dissipative and the asymptotic time evolution is the state of rest at the origin. The system looses its initial kinetic energy and finally stops. The main difference between this large time asymptotic behaviour and the previous one is that it is the same for all initial conditions. No matter where we start in phase space, if we wait long enough the trajectory will be very near to the origin. This is still a rather uninteresting large time behaviour but which is nevertheless met in many cases. For example if we put a viscous liquid in motion (steering a cup of coffee) and then stop the external action, the fluid will come to rest because of the viscous dissipation of the kinetic energy. It is also easy to produce more interesting time behaviour. One simply has to compensate the dissipative losses by some energy flux. As we will see, the gain and losses of energy may eventually equilibrate to produce non trivial asymptotic time dynamics. In the case of the damped harmonic oscillator, it is well known that if we add some harmonic forcing, the asymptotic motion in phase space will lie on a given ellipse. This will be true for all initial conditions. This case is however slightly different from our main study because the forcing is non autonomous. In many situations the forcing of energy is due to non-linear terms.

In the above discussion we have been considering mostly dissipative systems, and their study will be the main goal of these lectures. It should be mentioned that many of the ideas results and techniques presented below apply as well to the study of conservative systems.

Up to now we have been assuming that the time evolution of a system is described by a vector field on some finite dimensional space. Strictly speaking this is not the case for the Navier Stokes equation or for the Rayleigh Benard experiment. In this experiment a container filled with fluid is heated from below. For a small heat flux (which corresponds to a small energy forcing), the heat is simply conducted from the bottom to the top of the container through the fluid which remains at rest. There is however a threshold in the heat flux above which the fluid acquires a non zero velocity. The heat is then conducted but also convected to the top by the motion of the fluid. Slightly above the threshold the motion of the fluid is simple, one observes the formation of horizontal rolls (at least in small container). If one continues to increase the heating, other thresholds appear above which the motion of the fluid becomes more and more complex and eventually turbulent. This is a typical situation: at small energy forcing the dissipation wins and the asymptotic state is a state of rest. At well defined thresholds of the energy flux, the asymptotic dynamical behaviour changes qualitatively (bifurcations) and becomes more and more complex leading eventually to a chaotic behaviour. Note however that the parameter dependence of the system is non linear and therefore can be non monotone. This leads sometimes to the observation of a simplification (reverse bifurcation for example) of the dynamical state instead of an increase in the complexity.

In the above case the time evolution is described by a system of non linear partial differential equations. We will assume that the motion of the fluid is taking place in a finite container and we can therefore decompose the relevant fields (velocity, temperature etc.) using an adequate basis of functions adapted to the geometry of the container and the boundary conditions (Fourier series for example). The time evolution is of course reflected in the time dependence of the coefficients of this expansion. Introducing this decomposition into the evolution equation we get an infinite system of coupled non linear differential equations. In other words the system has an infinite number of degrees of freedom. It is natural at this point to ask wether all these modes are relevant to the problem and if one can recover the finite dimensional situation by considering only suitably chosen modes.

This turns out to be the case in many physical situations. Let us consider for example the Navier Stokes equation in dimension 2. The dissipative part is a Laplacian and if we go to Fourier space we notice that Fourier modes of large Fourier index will get a larger damping than those with small Fourier index. This leads to the intuition that Fourier modes of large index will evolve very rapidly and relax to a local (in time) equilibrium which changes in time due to the slower motion of the modes with small Fourier indices. One can say that the modes with large Fourier indices are slaves of the modes with small indices, and the system has only finitely many degrees of freedom excited. When the full non linearities are taken into account the problem becomes much more complex. In particular one should not expect that the natural modes are as simple as Fourier modes. The non linear theory has been developed rigorously in a number of interesting cases [FST]. In section 5 we will see some recently developed experimental techniques for measuring the number of relevant modes. There are now many experimental situations where a chaotic behaviour occurs with only a finite (small) number of degrees of freedom excited. One should however notice that the number of modes which is needed to describe the system increases in general with the energy forcing. The system may then reach a situation where the number of relevant modes is quite large, and the dynamical system approach becomes less accurate. In this situation often called fully developed turbulence, one should probably switch to a statistical description of the system (see [Li] for an experiment describing the transition, and [RS] for a theoretical review).

A simpler (but less satisfactory approach) to reduce the difficulty in the study of an infinite dimensional vector field is simply to truncate the infinite system to a finite one by artificially setting to zero an infinite number of components which are so strongly damped that it is reasonable to consider that they are negligible. The remaining modes are of course selected for their physical relevance ([Bh], [BPV]). Several very interesting and important finite dimensional systems were obtained in this way (Lorenz model [L], Henon map [H], etc.). It is obvious that these simplified models cannot reproduce accurately the experimental observations, however they still keep many of the interesting qualitative properties of the more realistic systems mostly because they are more or less normal forms (see chapter 2). These simplified models are also very important examples which are used in the development of the concepts in dynamical systems theory, with the idea that the concepts developed on simplified models can also be applied fruitfully in more realistic situations. Note also that some physical systems are indeed completely described by a finite number of degrees of freedom. There are many examples in mechanics, electronics etc.

In section 2 we will introduce the general notion of attractor which is the set where the asymptotic dynamics takes place. Then we will describe the elementary bifurcations which describe the transitions from the state of rest to the simplest non trivial dynamics. As we have explained above sequences of bifurcations will lead to more and more complex dynamics, and eventually to chaos. A particularly important such sequence (the period doubling road to chaos) will be studied in section 3. In section 4 we will describe some of the theoretical results about chaotic behaviour, and in section 5 we will discuss some of the recently discovered experimental techniques which are used to obtain information on the dynamics of a system from a time signal.

2. Stability and Elementary Bifurcations

As explained in the previous section we will be mostly interested by the dynamical behaviour for large time. This is not to say that all transient behaviours are unimportant, some of them are, but in general this is a more difficult problem and in fact some of the ideas discussed below can be useful in this situation.

If we come back to the simple example of the damped harmonic oscillator we see that for any initial condition the asymptotic dynamic is the same : it is a state of rest. This leads us to the general notion of attractor. An attractor is a compact subset Ω of phase space which is invariant by the time evolution and such that there is a neighborhood U which is also invariant by the evolution but contracts to Ω. More precisely, for any initial condition in U, the orbit converges to Ω. It is often useful to impose an irreducibility condition on the attractor, namely that it contains a dense orbit (this will avoid trivial complications). Note also that this notion is inadequate for conservative systems or for systems which preserve the volume of phase space.

One would like of course to determine the largest possible attracted neighborhood U of an attractor Ω. This is called the basin of attraction and it consists of all the initial condition giving rise to a trajectory converging to Ω. The global picture is therefore that in the phase space we have the basins of attraction of the attractors which are disjoint sets, but the complement is also an invariant set which can have a rather rich structure. For example, the boundaries of the various basins is an invariant set which can have a very complicated dynamics, although this is usually a thin and unstable set. Note also that if we have several attractors the large time dynamics of an initial condition will depend in which basin of attraction it is situated.

The simplest non empty set is a point, and it is therefore natural to ask when a point is an attractor. First it should be an invariant set (a fixed point, or stationary point) which means that the vector field should be zero at this point. It corresponds to the situation where the system is at rest (stationary state), and is met usually for low energy forcing as we have seen previously. For a point to be an attractor two conditions have to be satisfied. Assume for example that the time evolution is given by a vector field :

$$\partial_t x = X(x) ,$$

then if x_0 is an attractor it must be a fixed point, namely we must have

$$X(x_0) = 0 .$$

After one has solved the above equation for fixed points one has to check for the attractivity condition. If one starts with an initial condition $x_0 + \delta x(0)$ with $\delta x(0)$ small, one gets (by continuity) at time t a point $x_0 + \delta x(t)$, and the equation for $\delta x(t)$ is easily seen to be

$$\partial_t \delta x(t) = DX_{x_0} \delta x(t)$$

as long as $\delta x(t)$ is small enough so that higher order terms can be neglected. One then deduces easily the following theorem.

Theorem. If all the eigenvalues of the matrix DX_{x_0} are strictly inside the left half plane, then the fixed point x_0 is an attractor.

Note that the matrix DX_{x_0} has real coefficients, if we have a real time evolution. Therefore all the eigenvalues are either real or they come in complex conjugate pairs. It is also easy to prove that if one eigenvalue is in the right half plane the fixed point x_0 is not an attractor. The marginal situation where all eigenvalues are in the closed left half plane with one at least on the imaginary axis has to be investigated more carefully. We shall come back to this situation later when discussing bifurcation theory. Note also that the stability of a fixed point is a notion which is invariant by nonlinear changes of coordinates.

If one considers the case of discrete time iteration

$$x_{n+1} = T(x_n) ,$$

a similar analysis can be performed. A point x_0 is an attractor if first it is a fixed point (invariant set) :

$$T(x_0) = x_0 ,$$

and second if it is stable : the spectrum of the matrix

$$DT_{x_0}$$

must be strictly inside the unit disk. One can make the same remarks about the marginal situation as in the continuous time case.

We will now introduce a parameter dependence in the time evolution. As explained in the previous section, this parameter can be for example the amount of energy injected per unit of time in the system to compensate for dissipation. This is a very frequent physical situation, for example the Reynolds number in hydrodynamics measures the dissipation in the system (negative flux of energy). There is also a methodological reason to introduce a parameter dependence. The goal will be to study systems of increasing complexity starting with the simplest ones and trying to understand how more and more complex behaviours will develop. Ideally this complexity will grow when the parameter is increased. It turns out that the dependence in the parameter involves non linear effects which can sometimes break the monotonicity of the increase of the complexity, but the idea is nevertheless fully relevant for theoretical, numerical and experimental investigations. We will see how it leads to the study of bifurcations and scenarios (roads to chaos).

If for some parameter we have a stable fixed point for attractor, this situation will remain for nearby values of the parameter since the eigenvalues of the tangent matrix will change continuously (we are assuming here that the system has enough differentiability). Note however that in general the position of a fixed point varies with the parameter, and also the eigenvalues of the differential of the vector field.

For some value of the parameter, eigenvalues may reach the imaginary line (unit circle for iterations). Beyond that value the fixed point may still exist but becomes unstable (if one eigenvalue has a positive real part, respectively a modulus larger than one). Nearby initial conditions will flow away, eventually toward an other attractor. This change in the topology of the attractor is called a bifurcation. We have just seen that a stable fixed point can bifurcate only for values of the parameter for which an eigenvalue crosses the imaginary line (the unit circle for maps). Note also that a phase space may contain different attractors which can evolve when varying a parameter more or less independently, interact, collapse etc. This may lead for example to the hysteresis phenomenon.

One would like of course to classify the various possible bifurcations that can occur. If we consider the case of iterations, there are three possibilities for an eigenvalue to cross the unit circle. This is due to the already mentioned fact that eigenvalues are real or come in complex conjugate pairs. These three possibilities are as follows

1) One eigenvalue crosses at $+1$ (saddle-node bifurcation).
2) One eigenvalue crosses at -1 (pitchfork or period doubling).
3) Two complex conjugate eigenvalues cross together the unit circle at an angle which is neither 0 nor π (Hopf bifurcation).

One may of course observe that more complicated situations are possible (two eigenvalues simultaneously at $+1$, one at $+1$ and one at -1, etc.). All these situations correspond to adding supplementary constraints on the system, and (as is well known from linear analysis) can only be realized by adding other parameters. Such degenerate situations are not stable, they will split into the three elementary ones by small general perturbations. It can be proven that a bifurcation of a generic one parameter family will be one of the above three, namely typical systems, robust against perturbations will encounter only one of the three elementary bifurcations described above. A similar analysis holds for the case of continuous time evolution at the first crossing of the imaginary axis. We will describe below in more details the generic bifurcations. Note however that non generic bifurcations can be relevant in some special situations for example if the system possesses some symmetry which is not broken (or only partially) by the bifurcation. We refer to the literature [R] for these questions.

As mentioned earlier, one can reduce in some sense the number of degrees of freedom needed to describe the system. In the case of bifurcations this can be done completely and rigorously using the center manifold theorem [Sa], [GH], [R]. Intuitively, only the modes which correspond to direction(s) of the eigenvalue(s) which is (are) becoming unstable has (have) to be taken into account. The other directions remain stable and contract to the bifurcating mode(s). Note however that the rigorous statement is of nonlinear nature, it involves a nonlinear change of variables in phase space.

This result simplifies enormously the study of the elementary bifurcations. One can still try to simplify the study by using change of variables in the reduced phase space, this will eventually lead to a typical system called a normal form which is simple enough to be studied directly. We shall now describe briefly each situation using this normal form. It can be shown that under weak hypothesis the simple behaviours which are described below do occur generically. For a more detailed treatments we refer the reader to [A], [I], [GH], [IJ], [R] or [Sa]. In particular, one can give necessary and sufficient conditions expressed in terms of the various derivatives of the map at the fixed point for the occurrence of one of the particular bifurcation (note that this is better than a generic result).

1. *The Saddle Node Bifurcation.* In this case one eigenvalue crosses the unit circle at +1. The normal form is the one dimensional map (often called a logistic map)

$$x \to x^2 + 1/4 + \mu \, .$$

For $\mu < 0$ the system has a stable and an unstable fixed point. For $\mu = 0$, there is only one fixed point at $x = 1/2$, it has slope +1. This fixed point is stable for initial conditions $x_0 < 1/2$ and unstable for $x_0 > 1/2$. For $\mu > 0$ the fixed point has disappeared. In other words the bifurcation consists into the collapse of an attracting point with an unstable fixed point which annihilate each other (in this particular case they move to complex fixed points). What is left (for small μ) is a slowing down of the dynamics in the region of phase space where the fixed points collapsed. This phenomenon is known as intermittency [MP] and can be observed experimentally. The signal shows bursts of turbulence separated by intervals of quietness. These intervals of so called laminar behaviour are longer and longer if one approaches the critical value from above. One can, make quantitative predictions about the length of these quiet periods.

2. *Pitchfork (Period-Doubling or Flip) Bifurcation.* One eigenvalue crosses the unit circle at −1. The normal form is again one dimensional:

$$x \to g_\mu(x) = -(1+\mu)x - x^3 \, .$$

For $-2 < \mu < 0$, $x = 0$ is a stable fixed point. For $\mu = 0$ it has slope −1, and it is unstable from both sides. Looking at the double time map $g_\mu \circ g_\mu$, it is easy to see that for $\mu > 0$ the system still has a fixed point at $x = 0$ but which is now unstable. It has however a stable period two, namely a pair of points x_1 and x_2 such that

$$g_\mu(x_1) = x_2 \, , \; g_\mu(x_2) = x_1 \, .$$

In other words x_1 and x_2 are also fixed points of the map $g_\mu \circ g_\mu$. This period 2 is stable. Namely if one starts with an initial condition y_1 very near to x_1, the trajectory approaches the period in the sense that $g_\mu^{2n}(y_1)$ approaches x_1 and $g_\mu^{2n+1}(y_1)$ approaches x_2, where g_μ^p means the pth iterate of the map g_μ.

The pitch-fork bifurcation is therefore the appearance of an attractor which is a (stable) periodic orbit of period 2 emanating from a fixed point which becomes unstable. Note that contrary to the previous situation, the fixed point is still present for $\mu > 0$. This is a notable general fact. In a system which depends on a parameter, a fixed point stable or not can disappear only if one eigenvalue of the tangent map reaches the value $+1$ (0 for a vector field as we will see below). Note that this is a necessary but not sufficient condition.

This example of the pitch-fork bifurcation illustrates precisely the increase of the complexity of the asymptotic dynamic when the parameter is varied.

3. *Hopf Bifurcation.* This bifurcation occurs when a pair of complex conjugate eigenvalues cross the unit circle. In a generic situation, we can assume that none of the two bifurcating eigenvalues is a root of unity of degree less than or equal to 5 (note that we have already treated the cases $+1$ and -1). There are two normal forms. The first one is given by

$$T_\mu \begin{pmatrix} \rho \\ \theta \end{pmatrix} = \begin{pmatrix} (1+\mu)\rho - \rho^2 \\ \theta + \omega \mod 2\pi \end{pmatrix}$$

where ω is an irrational number. It is easy to see that for $\mu < 0$, $\rho = 0$ is a stable fixed point. For $\mu > 0$ it is unstable, and the attractor is now the stable limit cycle $\rho = \sqrt{\mu}$. Here, stable limit cycle means that nearby initial conditions will give rise to trajectories which spiral toward the limit cycle. One can show that for the generic Hopf bifurcation a stable limit cycle of radius $\mathcal{O}(\sqrt{\mu})$ emerges from the fixed point. The motion on the limit cycle can however be of various types depending on the so called rotation number. We refer to [A] for more details.

The second normal form is given by

$$T_\mu \begin{pmatrix} \rho \\ \theta \end{pmatrix} = \begin{pmatrix} (1+\mu)\rho + \rho^2 \\ \theta + \omega \mod 2\pi \end{pmatrix}$$

where again ω is irrational. For $\mu < 0$ we have a stable fixed point $\rho = 0$ and an unstable limit cycle $\rho = \sqrt{\mu}$. These two collapse at $\mu = 0$, and for $\mu > 0$ we are only left with an unstable fixed point at the origin. All other initial conditions will flow to infinity.

One should also mention that it is also possible to observe the inverse bifurcations by decreasing the parameter. As was mentioned earlier, these inverted bifurcations can even appear by themselves due to the fact that the parameter dependence may be non linear and non monotonous.

The case of vector fields is slightly easier to discuss. There are only two generic possibilities for the bifurcations. Either a simple real eigenvalue becomes positive and we have a saddle node bifurcation, or a complex conjugate pair crosses the imaginary axis (but not at the origin) and we have a Hopf bifurcation. The results are similar to those for maps : the saddle node case corresponds to the collapse of a stable and an unstable fixed point which annihilate each other, while for the Hopf bifurcation there are two cases. In one case a limit cycle (stable closed periodic orbit) emerges from the fixed point, or an unstable limit cycle coalesce at the fixed point. Note that in the case of vector fields there are no strong resonances for the Hopf bifurcation.

Using the method of reduction of vector fields to the iteration of maps one can discuss other simple cases. For example if one considers the Poincaré return map, one can discuss the stability and bifurcations of the associated limit cycle. Generic bifurcations will therefore consist in saddle node of limit cycles, doubling (the new attractor is also a limit cycle which closes after wrapping twice), or Hopf of a limit cycle associated with invariant tori.

As the parameter is further varied, more complicated bifurcations of the attractor can take place, and also interactions between attractors can occur. Reversed bifurcations can appear as well as hysteresis.

Trying to classify the various possibilities has led to the notion of road to chaos [Ec]. It is a sequence of bifurcations leading to chaotic dynamical states. One of the earliest road described (and of major conceptual importance) is the Ruelle-Takens-Newhouse road. For vector fields it starts with a fixed point which produces a limit cycle by a Hopf bifurcation. This limit cycle has then a secondary Hopf bifurcation resulting in a toral attractor. The result of Newhouse Ruelle and Takens is that the next bifurcation (destabilization of the torus) can lead to chaotic dynamics. As mentioned previously, several roads to chaos can take place independently and simultaneously in different parts of phase space. Initial conditions may select one or another. In the next section we will give some details on the road to chaos which is best understood.

3. The Period Doubling Scenario

The period doubling road to turbulence is at the moment the best understood, and also the most frequently observed (one reason is that it is stable against small perturbations). The phenomenology is very simple and can be explained easily using the one parameter family of maps of the interval (sometimes called logistic maps)

$$x \to g_\mu(x) = 1 - \mu x^2 \quad \text{for} \quad x \in [-1, 1] .$$

For $\mu < 3/4$, the attractor is a stable fixed point. At $\mu = \mu_1 = 3/4$ a period doubling (pitch-fork) bifurcation takes place, and as explained before, for μ slightly larger than μ_1, the attractor is a periodic orbit of period two (x_1, x_2). It is now natural to observe the system every other time step (stroboscope of period two). The time evolution is now given by $g_\mu^2 = g_\mu \circ g_\mu$, and the points x_1 and x_2 are now stable fixed points for this doubled period evolution. It is now easy to understand that increasing μ one may reach another period doubling bifurcation point μ_2 for the doubled period evolution. Coming back to the dynamics of the initial map g_μ, when μ crosses μ_2 the attractor bifurcates from a stable period 2 to a stable period 4. The new attractor is composed of four points which form a stable cycle of period 4. For the case of logistic maps one can prove that this process continues for ever. There is a third bifurcation point μ_3 where the attractor bifurcates from a stable period 4 to a stable period 8 etc. A striking property of this sequence of bifurcations is the asymptotic behaviour of the sequence $(\mu_n)_{n \in \mathbf{N}}$. It was first observed numerically [F] [CT] that

$$\frac{\mu_{n+1} - \mu_n}{\mu_n - \mu_{n-1}} \to \delta^{-1}$$

where $\delta = 4.6691\ldots$. In other words

$$\mu_\infty - \mu_n \sim \text{Cte}\, \delta^{-n} .$$

It turns out that many systems which depend on a parameter show this infinite sequence of period doubling bifurcations: Henon map, Lorenz system, truncated Navier-Stokes equations etc. Also this phenomenon is observed in experiments. What is even more surprising is that for all these systems the sequence of bifurcation value is asymptotically a geometric sequence and the same value of the rate δ is observed (universality). It should be emphasized that for these different situations the sequence of bifurcation values μ_n are different. It is only the asymptotic geometrical rate of these different sequences of numbers which is universal. Other universal numbers associated to this accumulation of period doubling have been described. For example there is a universal scaling factor $\lambda \simeq .39\ldots$, for the logistic map it is the scaling factor for the distance between the critical point of the map $(x = 0)$ and the nearest point of the periodic attractor. More recently, the dimension spectrum (describing the attractor at the accumulation of period doubling) was also shown to be universal (see [CLP] for more details and references).

Similar universal behaviours are also known to occur in second order phase transitions where they are explained using renormalization group analysis. The same approach was used to explain the above universality. One builds first a renormalization transformation $\mathcal{N}$. It is based on the ideas explained above. First one doubles the unit of time and then changes the scale of the problem. After these transformations the dynamic looks roughly the same. For a mapping f of the interval $[-1,1]$ like a logistic map, the explicit formula for the renormalization transformation $\mathcal{N}$ is then given by

$$\mathcal{N}(f)(x) = \frac{1}{f(1)} f \circ f(f(1)x) ,$$

where we have assumed that f is even, concave with only one critical point at $x = 0$ and with critical value $f(0) = 1$. The scaling is chosen to maintain these normalizations. Doubling the unit of time is obtained by composition.

It turns out that $\mathcal{N}$ has a fixed point with an unstable space of dimension 1. The number δ is the expanding coefficient at this fixed point (the simple eigenvalue of the tangent map at the fixed point which has a modulus larger than one). It corresponds to the exponential of a critical index in the language of second order phase transitions. One parameter families of time evolutions are equivalent to temperature trajectories. One can now apply the methods of scaling limit to analyze the transition and to prove the above results. A practical way to do this is to introduce the bifurcation manifold $\mathcal{B}$, this is the set of maps f normalized as above and with the property that if x_f is the fixed point of f, then $f'(x_f) = -1$. From the discussion of the previous section we conclude that when a one parameter family f_μ of maps crosses $\mathcal{B}$ transversally, there is a period doubling bifurcation. It is now easy to see that the number μ_n where a stable period 2^{n-1} bifurcates to a stable period 2^n is the value of μ where the curve $\mathcal{N}^n(f_\mu)$ crosses $\mathcal{B}$. The above results: geometrical accumulation for the sequence of period doubling, universality etc. follow now by simple geometrical considerations as in the case of second order phase transitions. In particular, universality comes from the fact that the map $\mathcal{N}$ is independent of a particular one parameter family. In other words this map it is universal, and all the associated quantities are universal. We refer to [CE], [EE], [E] and [Su] for more details.

These ideas were developed recently for other dynamical systems like critical circle maps, area preserving maps, hamiltonian systems etc. As in the theory of phase transitions, one has to introduce the notion of universality classes which contain all the problems with the same universal indices. The value of these indices varies from one class of problems to another, although the form of the universality results remain the same. In the above situation of maps of the interval, the nature of the critical point (quadratic, quartic etc.) can change the universality class.

4. Chaotic Behaviour

We have seen before that changing a parameter in a nonlinear system can lead to bifurcations of attractors. In this way one may get attractors of more and more complex geometry supporting more and more complex time evolutions. Simple attractors like points, closed curves, are in general associated to simple dynamics. Chaotic dynamics is usually associated with attractors of complicated geometry which have non integer (Hausdorff) dimension. These so called strange attractors emerge from successive (more and more complex) bifurcations as the parameter is varied. This is for example the case at the accumulation of period doubling for the one parameter family $\mu \to 1 - \mu x^2$. One can prove that at the end of the infinite sequence of period doubling bifurcations the attractor is a Cantor set of dimension .56.... There are now many known examples of dynamical systems with a strange attractor. One of the simplest case is the Lorenz vector field (or more accurately the geometrical Lorenz flows) [GH], [R]. The solenoid is a nice example for discrete time [G], [R]. It has the advantage of being rigorously understood, whereas this is not the case for the most "popular" (and theoretically most important) Henon system. See however [BC] for recent results.

It has been known for a long time in ergodic theory that dynamical systems may present different qualities of chaos. This can be easily shown by considering two different dynamical systems on the circle.

The first example is the irrational rotation given by

$$\theta \to \theta + \omega \quad \mod 1 ,$$

where ω is an irrational number. This transformation preserves the Lebesgue measure which is ergodic. However this dynamical system is not very chaotic. For example, if one consider two nearby initial conditions, it is obvious from the explicit formula that the distance between the nth iterates remains forever of the same order of magnitude (in particular no mixing occurs).

Consider on the other hand the transformation of the unit circle given by

$$\theta \to 2\theta \quad \mod 1 .$$

One can easily show that this map also preserves the Lebesgue measure which is ergodic (since the map is not invertible, invariance of the Lebesgue measure means that the measure of a measurable set is equal to the measure of it's set theoretic pre-image). However, if one considers two nearby initial conditions in generic position, it is an easy exercise to show that the distance between the iterates grows exponentially fast (as long as it remains a small number).

The analysis of these qualitatively very different time behaviours leads naturally to a qualitative definition of strong chaos which is called sensitive dependence to initial condition (s.d.i.c.). Let T be a discrete time evolution with an attractor Ω. T has s.d.i.c. if for every $\varepsilon > 0$, and every initial condition $x_0 \in \Omega$, there is another initial condition y_0 in Ω such that $d(x_0, y_0) < \varepsilon$, and a number n (which may depend on x_0, y_0 and ε) which satisfy

$$d(T^n(x_0), T^n(y_0)) = \mathcal{O}(1) .$$

This means that on the attractor some small errors will be dramatically amplified by the time evolution. It is obvious that this is not the case for the irrational rotation on the circle.

This idea was nicely illustrated by E. Lorenz for the dynamical system associated to the motion of the atmosphere (General Circulation). He argued that if a butterfly of California flaps its wings, after a period of about two weeks the dynamical state of the atmosphere in Valparaiso will be different from what it would have been if the butterfly had not moved. In other words s.d.i.c. prevents from making long time predictions on the system.

It is natural at this point to ask if one can give a more quantitative approach to s.d.i.c. This leads to the notion of Lyapunov exponents. This notion is a direct generalization of the stability analysis of fixed points (periodic orbits) described before. For simplicity we start with a one dimensional map f of the interval $[-1, 1]$. If x_0 and $y_0 = x_0 + h$ (h small) are two nearby initial conditions, we can ask for the distance between the points on the respective orbits after n time steps. We have obviously

$$f^n(x_0 + h) - f^n(x_0) \simeq h f^{n\prime}(x_0) ,$$

and this formula is valid as long as the left hand side of the equation remains small. We can now use the chain rule and we obtain

$$f^{n\prime}(x_0) = \prod_0^{n-1} f'(f^j(x_0)) = \prod_0^{n-1} f'(x_j)$$

if we denote by x_j the point $f^j(x_0)$. This formula expresses the growth of the distance as a product of n terms and it is therefore natural to test for the exponential growth in n of $|f^{n\prime}(x_0)|$. This is done by considering the quantity

$$\frac{1}{n} \sum_0^{n-1} \log |f'(x_j)|$$

which is nothing but the time average along the trajectory of x_0.

If the limit exists when $n \to \infty$ it is called the Lyapunov exponent of the orbit of x_0. Assume now that ν is an ergodic invariant probability for f. Then it follows from ergodicity and from the ergodic theorem that (if $|\log|f'||$ is ν integrable) for ν almost every initial condition the Lyapunov exponent exists and is (ν almost surely) independent of the initial condition. One has however to be careful that the Lyapunov exponent usually depends on the ergodic measure ν.

In dimension larger than one the analysis is more delicate and the result is due to Oseledec and Ruelle. For a given transformation T and an initial point x_0, one starts the argument as before. However the problem is now to understand the large n behaviour of the norm of the vector

$$DT_{T^{n-1}(x_0)}\cdots DT_{x_0}h$$

(h is a vector). The difficulty is that in general the matrices in the above product do not commute. In order to understand the general result, let us consider the particular situation where x_0 is a fixed point. We then have the power of a fixed matrix. Let us moreover assume that this matrix is diagonal with eigenvalues 2 and 1/2. In other words we want to know how large is the vector

$$\begin{pmatrix} 2 & 0 \\ 0 & 1/2 \end{pmatrix}^n \begin{pmatrix} h_1 \\ h_2 \end{pmatrix} .$$

The answer is now easy. If $h_1 \neq 0$ the norm of this vector for n large enough grows exponentially fast with an exponential rate equal to $\log 2$. On the other hand, if $h_1 = 0$ (but $h_2 \neq 0$) the norm decreases exponentially fast with a rate equal to $-\log 2$. In the general case, one has again to choose first an ergodic invariant measure ν. It can be proven under some mild conditions that for ν almost every initial condition x_0, the sequence of positive matrices

$$\Lambda_n = \left(DT^{n\,t}_{x_0} DT^n_{x_0}\right)^{1/2n}$$

has a limit when $n \to +\infty$. The Lyapunov exponents $\lambda_0 > \lambda_1 > \lambda_2 > \cdots$ are the logarithms of the eigenvalues of the limiting matrix. Moreover, there is a sequence of linear spaces $E^1_x \supset E^2_x \cdots$ defined for ν almost every x and such that

1) $DT_x E^j_x = E^j_{T(x)}$,
2) E^1_x is the tangent space to the phase space,
3) $\lim_{n\to\infty} n^{-1} \log \| DT^n_x h \| = \lambda_j$ for all $h \in E^j_x - E^{j+1}_x$.

There is also a non linear version of this result which asserts the existence of local invariant manifolds, we refer to [GH] and [R] for more details.

Note that for dissipative systems the sum of the Lyapunov exponents must be negative (this is the exponential rate of contraction of the volume). If on the other hand the mapping is volume preserving, the sum of the Lyapunov exponents is equal to zero.

If the attractor is a stable fixed point, it follows from the definition that all Lyapunov exponents are negative. An attractor with at least one positive Lyapunov exponent is called hyperbolic. As explained before, one has to refer to an ergodic invariant measure in this definition. This is a difficulty because in general strange attractors support many ergodic invariant measures. A natural choice of measure was proposed by Ruelle and Bowen. For a transformation T attracting an open set U to an attractor Ω this measure ν is defined by requiring that

$$\frac{1}{n}\sum_{0}^{n-1} g(T^j(M_0)) \to \int_{\Omega} g(M)d\nu(M) \ .$$

for every function g continuous on U (observable) and for Lebesgue almost every initial condition M_0 in U. Note that the above formula is different from the usual ergodic theorem because ν is in general singular with respect to the Lebesgue measure (Ω may be of Lebesgue measure zero). In other words the time average of an observable is equal to a (weighted) space average on the attractor, a consequence of dissipation (contraction of phase space). This special measure is known to exist in some interesting cases (the so called axiom A systems), where it has some other interesting properties (like the stability with respect to noise). It is called an S.R.B. measure (for Sinai, Bowen, Ruelle) if it satisfies some more technical assumptions (see [B] and [R] for details). Note also that for the good cases there is a dissipative analog to the Gibbs ansatz for constructing this S.R.B. measure.

It may seem at first that the above notion of hyperbolicity (stretching in at least one direction) contradicts the notion of attractor and in particular the compactness assumption. This is not so if we consider nonlinear systems, and the combination of these two properties (instability and compactness) is indeed at the origin of chaos. To understand this fact more precisely, consider a small segment on the attractor pointing in an unstable direction (i.e. a direction with positive Lyapunov exponent). The time evolution will stretch this segment exponentially fast. When this segment has been expanded to a macroscopic size the non linearities will fold it. This folded segment will be expanded again then folded and so on. After a large number of iterations, the curve will look very complicated (chaos) and eventually will be dense in the attractor (this is related to ergodicity). Therefore we see that the combination of instability and compactness is responsible for the complexity of the dynamics and of the geometry of the strange attractors. This argument has been developed rigorously by S. Smale [S] who introduced the notion of horseshoe, and their relation with homoclinic points (transverse crossing of the stable and unstable invariant manifolds of a periodic point). It was shown later on [K] that if a system has positive entropy (see below for a definition), then homoclinic points (and horseshoes) are dense in the attractor. The converse of this result is false: if a system has a homoclinic point, it has chaotic trajectories but the S.R.B. measure may nevertheless be associated to a regular motion (there are many such examples for maps of the interval where the attractor is a stable periodic orbit). An earlier consequence of the presence of homoclinic points is due to Poincaré who showed that if there is a homoclinic point, there cannot exist an analytic integral of motion [A]. On the other hand it is in general difficult to prove that there is a positive Lyapunov exponent for an S.R.B. measure. The problem of deciding if there is a homoclinic point is simpler and there is in particular the important technique of Melnikov (see [GH]). We stress again however that this is not enough to prove that there is strong chaos. It may occur that the system has complicated trajectories in a subset of phase space of zero lebesgue measure and the typical asymptotic dynamic corresponds to a simple attractor. In the last chapter we will describe some experimental and numerical tests which are useful in the absence of a theoretical understanding of the dynamics.

Another consequence of the above analysis is a lower bound on the number of degrees of freedom which is needed to have hyperbolic chaos. For iterations one needs at least one unstable direction (s.d.i.c.) and one stable (attracting) direction transverse to the attractor, i.e. two degrees of freedom. Chaotic (Axiom A) two dimensional systems were constructed by Plykin [P]. In some sense this is also the case for the Henon attractor ([BC]). For flows one needs one more dimension (the time direction of the vector field), i.e. three dimensions. An example is given by the Lorenz model (or more rigorously by the geometrical Lorenz flows) [GH]. We also note that for non invertible maps, the stable direction is not needed, and one can have chaos with maps of the interval or of the circle. We already mentioned the simple example of the transformation

$$x \longrightarrow 2x \mod 1$$

for which the Lebesgue measure is ergodic and the Lyapunov exponent is $\log 2$.

Other numbers are used to describe quantitatively the intensity of chaos. We will briefly describe two important ones: the entropy and the dimension. We will start with the entropy. As is well known in Statistical Mechanics, the entropy measures the number of possible configurations. For dynamical systems, the entropy measures the number of distinct trajectories, where distinct means different for a given accuracy. This leads first to the definition of the topological entropy.

Let T be a homeomorphism of a compact metric space X (for example an attractor) with a distance d. For a given positive number ε (the accuracy) and an integer n we say that two trajectories corresponding to two different initial conditions x_0 and y_0 are (n, ε) different if there is a time j, $0 \leq j \leq n$ such that $d(T^j(x_0), T^j(y_0)) > \varepsilon$. In other words we can distinguish the trajectories at precision ε. Given n and ε as above, if x_0 and y_0 give rise to identical (n, ε) trajectories we shall say that they belong to the same (n, ε) class. Let $N(n, \varepsilon)$

denote the minimal number of (n, ε) classes covering X. The topological entropy of T, denoted by $h(T)$ is defined by

$$h(T) = \lim_{\varepsilon \to 0} \limsup_{n \to +\infty} \frac{1}{n} \log N(n, \varepsilon) .$$

If the topological entropy is positive, one expects many different trajectories, and therefore chaos. As we have seen before, this does not imply that the relevant measure (S.R.B. for example) is associated to chaotic motion. This chaos can take place for example on the boundary of the basins of different attractors. Note that if T is a diffeomorphism with a homoclinic point then it has positive topological entropy.

A more precise notion is the metric (also called Kolmogorov-Sinai) entropy of an invariant ergodic probability measure ν. We will not start with the standard definition of ergodic theory but with an equivalent one [K] which is related to the above definition of the topological entropy. The only difference with the previous definition is that one neglects some (untypical with respect to ν) initial conditions. For n and ε as above and for a positive number δ (smaller than one), we define $N(n, \varepsilon, \delta)$ to be the minimal number of (n, ε) classes needed to cover some set of measure larger than $1 - \delta$. The metric entropy of ν, $h(\nu)$ is defined by

$$h(\nu) = \lim_{\delta \to 0} \lim_{\varepsilon \to 0} \lim_{n \to +\infty} \frac{1}{n} \log N(n, \varepsilon, \delta) .$$

The metric entropy has many interesting properties. We shall only mention a relation with positive Lyapunov exponents:

$$h(\nu) = \sum_{\lambda_i > 0} \lambda_i .$$

This is known as Pesin's formula, and has been extended to all S.R.B. measures [Pe].

Positivity of the entropy of the S.R.B. measure is therefore a criteria for strong chaos. It was shown in [K] that it implies the existence of homoclinic points.

Another interesting number that can be extracted from dynamical systems is a dimension. The dimension of the attractor is not the most interesting dimension like the topological entropy is not the most interesting entropy. The relevant number is the dimension $\dim_{\mathrm{H}}(\nu)$ of the invariant ergodic probability ν (in general the S.R.B. measure). This number is the infimum of the Hausdorff dimensions of the sets of ν full measure. It is in general smaller than the Hausdorff dimension of the support of ν (which is contained in the attractor). There is another characterization of this dimension [Y]. Let $B_r(x)$ be the ball of radius r centered at a point x. If ν is an ergodic measure such that

$$\lim_{r\to 0} \frac{\log \nu(B_r(x))}{\log r}$$

exists ν a.s., then this number is the dimension of ν. This is easily verified if ν is the arclength of a curve or the area measure on a surface. L.S. Young [Y] showed moreover that for two dimensional compact hyperbolic attractors there is a relation between entropy, dimension and Lyapunov exponents given by

$$\dim_{\mathrm{H}}(\nu) = h(\nu)\left[\frac{1}{\lambda_1} + \frac{1}{|\lambda_2|}\right]$$

where $\lambda_1 > 0 > \lambda_2$ are the Lyapunov exponents. Another dimension, the Lyapunov dimension $\dim_\Lambda(\nu)$, was defined by Kaplan and Yorke. If $\lambda_1 \geq \lambda_2 \geq \cdots$ is the ordered sequence of Lyapunov exponents (with multiplicity), and if k is the smallest integer such that

$$\sum_{i=1}^{k+1} \lambda_i < 0,$$

we can define an affine function $c_\nu(s)$ by

$$c_\nu(s) = \sum_{i=1}^{k} \lambda_i + (s-k)\lambda_{k+1} \quad \text{for} \quad k \leq s < k+1\,.$$

The Lyapunov dimension $\dim_\Lambda \nu$ is defined by

$$\dim_\Lambda \nu = \sup\{s \,|\, c_\nu(s) \geq 0\}\,.$$

It has been proven that

$$\dim_{\mathrm{H}}(\nu) \leq \dim_\Lambda(\nu)\,.$$

We refer to the literature for some recent developments on partial dimensions and on the so called dimension spectrum. See for example [LY] and [CLP].

5. Analysis of Experiments

Some important recent developments in the theory of dynamical systems are devoted to the experimental measurements of some of the quantities introduced in the last section. The new methods and ideas apply to the output generated by computer simulations and to experimental signals as well.

One of the basic tool in analyzing a signal is to look at it's power spectrum. It is well known that discrete peaks indicate quasi periodic motion while broad band spectrum is related to chaos. However the power spectrum does not seem adequate for a precise analysis of chaotic behaviours. More precise information can be obtained from a direct analysis of the signal.

The direct analysis of a signal is based on a fundamental theorem of F.Takens [T]. Suppose we have a discrete time evolution T for a system with an attractor of dimension m (in this sense there are m relevant degrees of freedom). Then for a generic real observable g the theorem asserts that one can reconstruct the system (at least on the attractor) in a space of dimension $2m+1$ by looking at the points

$$(g(x), \cdots, g(T^{2m}(x)))$$

at successive instants of time. This remarkable result has different variants which have to be used according to the performed measurements. We refer to the literature for more details about the implementation of the algorithm (see [ER] and [BPV]). Note also that in modern technology most signals are digitalized and we shall not mention anymore the continuous time case. At this point, given an experimental signal, the number m is not known (it may be infinite in which case the theorem does not apply). We shall see below how it is determined. We emphasize that this reconstruction is done out of a one dimensional signal. The results are of course more accurate if one can obtain directly from the experiment a higher dimensional signal. Strictly speaking, the condition of genericity of the observable is impossible to verify, and should be replaced by the hypothesis that the signal satisfies this assumption (i.e. that obvious experimental flaws have been avoided). Note also that this method reconstructs the attractor and also the dynamics by looking at the time evolution of the observable.

Using the above reconstruction, one can try to determine the dimension of the S.R.B. measure. One fixes first a tentative number q for the reconstruction dimension (the number m above). Then for a given (reconstructed) trajectory $x_1, x_2, \cdots, x_N$ and for a positive number r, one computes the quantity

$$C_q(r) = \frac{1}{N^2} \sum_{i,j=1}^{N} \theta(r - d(X_i, X_j)) ,$$

where θ is the Heaviside function. This quantity is supposed to approximate the measure of a sphere of radius r (by the ergodic theorem), averaged over it's center (see the result of L.S.Young in the previous section which shows that the measure of a typical ball of radius r is $r^{\dim_H(\nu)}$). The next step is to plot the function $-\log C_q(r)$ versus $-\log r$ for r small. The graph has eventually a straight piece for moderately small values of r. That is to say for values of r which are small enough with respect to the size of the attractor but large enough so that a time average of length N gives a good approximation of the measure. One then plots the slope of this straight piece as a function of the reconstruction dimension q. For large enough q, the slope may tend to an asymptotic value which is the so called information dimension. We refer to [ER] for practical details. Note that the reconstruction dimension $2m+1$ which was unknown at the beginning is also obtained by this method. Strictly speaking the dimension which is obtained here is not the dimension of the measure (nor the dimension of the attractor). It is however a very useful information and in most systems it is a good approximation for the dimension of the S.R.B. measure (and for the dimension of the attractor).

Using the reconstruction described above, it is also possible to measure Lyapunov exponents. The first difficulty is to construct locally a good approximation of the tangent map. This can be done by finding the best linear fit for a cluster of nearby points and their images. When this is done along a trajectory, one can extract the Lyapunov exponents by successively triangulating the product of these tangent maps (see [EKRC] for details of implementation).

Some other quantities can be measured from the experimental (or numerical data). These include the entropy and more recently the singularity spectrum [JKL]. Using these different information together one can usually derive a precise picture of the successive bifurcations and of the nature of chaos in experiments governed by systems with few degrees of freedom excited. It should be stressed again that the idea of varying a natural parameter is of fundamental importance in both experimental and numerical investigations. We refer to [BPV] and [ER] for more details.

References

[A] Arnold, V., *Chapitres Supplémentaires de la Théorie des Equations Differentielles Ordinaires.* Mir, Moscou (1980).

[AA] Arnold, V.I., A. Avez, *Problèmes Ergodiques de la Mecanique Classique.* Gauthier-Villars, Paris (1967).

[B] Bowen, R., *Equilibrium States and the Ergodic Theory of Anosov Diffeomorphisms.* Lecture Notes in Mathematics 470, Springer-Verlag, Berlin, Heidelberg, New-York (1975).

[BC] Benedicks, M., L. Carleson, The Dynamics of the Hénon Map. Ann. of Math. (to appear).

[Bh] Bhattacharjee, J.K., *Convection and Chaos in Fluids.* World Scientific, Singapore (1987).

[BPV] Bergé, P., Y. Pomeau, C. Vidal, *Order Within Chaos.* Wiley-Hermann, Paris (1987). P.Bergé. *Le Chaos, Théorie et Expériences.* Eyrolles, Paris 1988.

[CE] Collet, P., J.-P. Eckmann, *Iterated Maps of the Interval as Dynamical Systems.* Birkhauser, Boston (1980).

[CLP] Collet, P., J. Lebowitz, A. Porzio, Journ. Stat. Phys. **47**, 609 (1987).

[CT] Coullet, P., C. Tresser, C. R. Acad. Sci. **287**, 577 (1978).

[Ep] Epstein, H., Commun. Math. Phys. **106**, 395 (1986). Non Linearity **2**, 305 (1989).

[Ec] Eckmann, J.-P., Rev. Mod. Phys. **53**, 643 (1981).

[EE] Epstein, H., J.-P. Eckmann, Commun. Math. Phys. **107**, 213 (1986).

[ER] Eckmann, J.-P., D. Ruelle, Rev. Mod. Phys. **57**, 617 (1986).

[EKRC] Eckmann, J.-P., S. Kamphorst, D. Ruelle, S. Ciliberto, Phys. Rev. **A34**, 4971 (1986).

[F] Feigenbaum, M., J. Stat. Phys. **19**, 25 (1979).

[FST] Foias, C., G. Sell, R. Temam, J. Diff. Equ. **73**, 309 (1988).

[GH] Guckenheimer, J., P. Holmes, *Nonlinear Oscillations, Dynamical Systems, and Bifurcations of Vector fields.* Springer-Verlag, Berlin, Heidelberg, New-York (1983).

[H] Hénon, M., Commun. Math. Phys. **50**, 69 (1976).

[I] Iooss, G., *Bifurcation of Maps and Applications.* Mathematical Studies Vol. 36, North Holland, Amsterdam (1979).

[IJ] Iooss, G., D.Joseph, *Elementary Stability and Bifurcation Theory.* Springer-Verlag, Berlin, Heidelberg, New-York (1981).

[JKL] Jensen, M., L. Kadanoff, A. Libchaber, Phys. Rev. Lett. **55**, 2798 (1985).

[K] Katok, A., Publ. Math. IHES **51**, 137 (1980).

[L] Lorenz, E.N., J. Atm. Sci. **20**, 130 (1963).

[Li] Libchaber, A. et al., J. Fluid Mech. **204**, 1 (1989).

[LY] Ledrappier, F., L.S. Young, Ann. of Math. **122**, 509 (1985).

[MP] Manneville, P., Y. Pomeau, Commun. Math. Phys. **74**, 189 (1980).

[P] Plykin, R., USSR Math. Sb. **23**, 233 (1974).

[Pe] Pesin, J.B., Russ. Math. Surv. **32**, 55 (1977).

[R] Ruelle, D., *Elements of Differentiable Dynamics and Bifurcation Theory.* Academic Press, London 1989. *Chaotic Evolutions and Strange Attractors.* Cambridge University Press, Cambridge 1989.

[RS] Rose, H., P. Sulem, J. Phys. **39**, 441 (1978).

[S] Smale, S., Bull. Amer. Math. Soc. **73**, 747 (1967).

[Sa] Sattinger, D.H., Bull. Amer. Math. Soc. **3**, 779 (1980).

[Su] Sullivan, D., On the Structure of Infinitely Many Dynamical Systems Nested Inside or Outside a Given One. Preprint IHES 1990.

[T] Takens, F., In *Dynamical Systems and Turbulence.* Lecture Notes in Mathematics Vol. 898, Springer-Verlag, Berlin, Heidelberg, New-York (1980).

[Y] Young, L.S., Ergod. Theor. & Dyn. Sys. **2**, 109 (1982).

SHOCKS IN THE BURGERS EQUATION AND THE ASYMMETRIC SIMPLE EXCLUSION PROCESS

P. A. FERRARI
Instituto de Matemática e Estatística
Universidade de São Paulo
Cx. Postal 20570
01498 São Paulo
Brasil

1. Introduction

The Burgers equation is used as a model of transport in highways. The function $u(r,t)$ represents the density of cars at $r \in \mathbb{R}$ at time t. We assume that, in the absence of other cars, the velocity of a single car is θ. Due to the presence of other cars this velocity can be lower. The variation of density at r in an infinitesimal time interval dt is given by the number of cars entering in the infinitesimal interval $(r, r+dr)$ that is $\theta u(r-dr,t)(1-u(r,t))$ minus the number of cars exiting that interval: $\theta u(r,t)(1-u(r+dr,t))$ (note that $1-u$ is the density of free space in the interval). Hence the density must satisfy the equation

$$\frac{\partial u}{\partial t} = -\theta \frac{\partial [u(1-u)]}{\partial r} \tag{1.1.a}$$

This is called the (unviscous) Burgers equation. We think of the highway as having only one lane. One of the most interesting phenomena in highways is the formation of shock waves. They form when for some reason one car lowers its velocity or stops. Then, after a small but positive time, the next car must lower its velocity and gets closer to the previous one. Then the third car lowers its velocity and so on. In this way the cars are divided into two well differentiated regions: one of high density where the cars are packed and run slowly and the other of fast cars and low density. This phenomena has been observed in actual highways by Walker [wa]. The shock at a given time is given by the car that is applying brakes at that time. To model the highway at a microscopic level –*i.e.* taking into account the interaction between single cars– a stochastic model called the asymmetric simple exclusion process has been used. The assumptions that we used to derive the Burgers equation (1.1.a) are rigorously proven for this model when the initial conditions are non decreasing [bfsv]. We study the case of non decreasing initial

E. Goles and S. Martínez (eds.), Statistical Physics, Automata Networks and Dynamical Systems, 25–64.

conditions that present only one shock: the initial condition will be constant to the right and left of the origin and present a discontinuity at the origin.

In the simple exclusion process, cars are represented by particles that sit at the integers. At most one particle is allowed at each site. Each particle waits an exponentially distributed random time of parameter θ and attempts to jump to the nearest right neighbor. The jump is actually realized only if the site is empty.

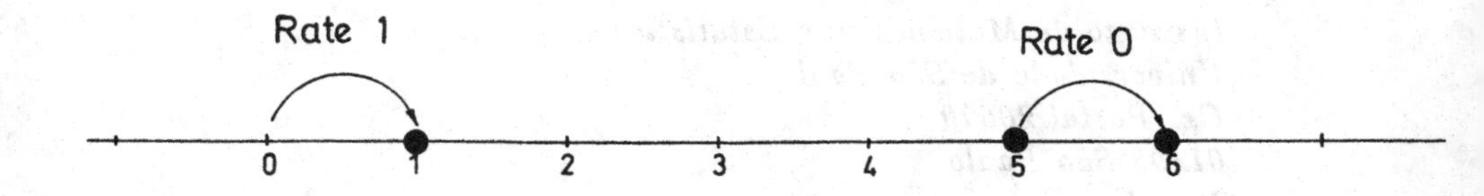

Figure 1. The simple exclusion process.

In Figure 1 we represent particles by black balls. The particle at the origin has rate θ for jumping because site 1 is empty. The particle at site 5 has rate zero as site 6 is occupied. As we will see, the essential properties of travelling waves of actual highways can be rigorously described using the simple exclusion process. The main feature of this process is the following: if one starts the system with a distribution that has densities ρ to the left, and λ to the right, $\rho < \lambda$, then there exists a random position that also obey local rules so that the system as seen from that position, uniformly in time, has densities λ to the right and ρ to the left, asymptotically.

On the way we review the proof of Liggett [l] that the set of all invariant measures is given by convex combinations of the translation invariant product measures and the blocking measures. The former are the measures ν_ρ, obtained by putting independently at each site a particle with probability ρ and no particle with probability $1 - \rho$. The blocking measures are the measures that concentrate mass on configurations with particles to the right of a given site and no particles to its left. Clearly these configurations are invariant.

The simple exclusion process has been introduced by Spitzer in [spi]. The set of invariant measures was described by Liggett [l]. The hydrodynamical limit has been studied by Rost [r], Andjel and Vares [av] and Benassi and Fouque [bf1]. The existence of a microscopic shock was studied in the case of vanishing left density by Ferrari [f1], Wick [w], De Masi, Kipnis, Presutti and Saada [dkps] and Gärtner and Presutti [gp]. In the case of non vanishing left density, the existence of a microscopic shock was simulated by Boldrighini, Cosimi, Frigio and Nunes [bcfg] and proven by Ferrari, Kipnis and Saada [fks]. The present approach follows Ferrari

[f2]. Bramson [b] and Lebowitz, Presutti and Spohn [lps] and Spohn [S] reviewed some of the results. Other related results are due to Kipnis [k] who proved a central limit theorem and law of large numbers for the position of a tagged particle and to De Masi and Ferrari [df] who computed the variance of the limiting Gaussian distribution.

2. The Burgers Equation

In this section we describe the (unviscous) Burgers equation and show how to find weak solutions when the initial condition is a shock. We briefly sketch the geometric approach of Lax [lax].

The inviscid Burgers equation is the hyperbolic equation (in what follows we take $\theta = 1$ in (1.1))

$$\frac{\partial u}{\partial t} = -\frac{\partial[u(1-u)]}{\partial r} \tag{2.1.a}$$

We consider the initial value problem $u(r,0) = u_0(r)$, where

$$u_0(r) = \rho 1\{r \le 0\} + \lambda 1\{r > 0\} \tag{2.1.b}$$

(shock initial conditions). The way to find the solutions is called the method of characteristics. If one calls $a(u) = \frac{\partial}{\partial u}(u(1-u)) = (1-2u)$, then the equation can be written

$$\frac{\partial u}{\partial t} + a(u)\frac{\partial u}{\partial r} = 0$$

so that u is constant along trajectories $w(r,t)$ with $w(r,0) = r$, that propagate with speed $a(u)$. These trajectories are called characteristics. They are straight lines and allow to construct a solution of the equation for t small. If different characteristics meet, giving two different values to the same point, then the solution develops a discontinuity. Ours is the simplest case, when the discontinuity is present in the initial condition. Indeed, for $r > 0$, the characteristics starting at r and $-r$ have speed $(1-2\lambda)$ and $(1-2\rho)$ respectively and meet at time $t(r) = r/(\lambda-\rho)$. Calling $f(u) = u(1-u)$ and using the conservation law of the equation it is not difficult to show that the discontinuity propagates at velocity $v := (f(\lambda)-f(\rho))/(\lambda-\rho) = 1-\lambda-\rho$.

In Figure 2 time is going down. We have drown the characteristics starting at r and $-r$ that go at velocity $1-2\lambda$ and $1-2\rho$ respectively. The center line is the shock that travels at velocity $1-\lambda-\rho$. The solution $u(r,t)$ is λ for $r > vt$ and ρ for $r < vt$ *i.e.* $u(r,t) = u_0(r-vt)$. This means that for all continuously differentiable test functions $\Phi(r,t)$,

$$\int\int \left(\frac{\partial \Phi}{\partial t}u + \frac{\partial \Phi}{\partial r}u(1-u)\right) drdt = 0$$

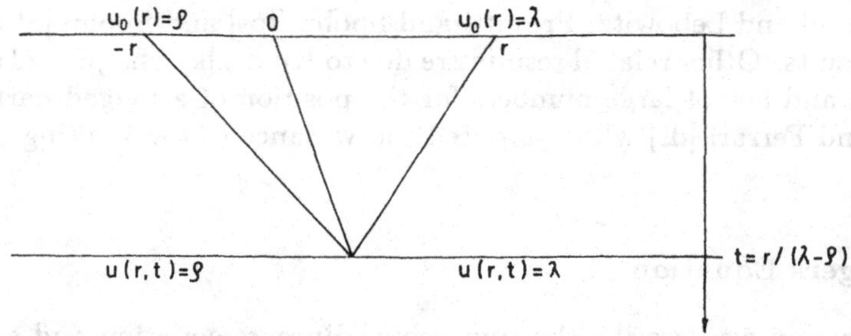

Figure 2. Shocks and characteristics in the Burgers equation.

Unfortunately the solution we gave above is not unique. But this is the one with physical interest because it comes as a limit when $b \to 0$ of the (unique) solution of the (viscous) Burgers equation

$$\frac{\partial u}{\partial t} = -\frac{\partial[u(1-u)]}{\partial r} + b\frac{\partial^2 u}{\partial r^2} \tag{2.2}$$

This solution is called entropic and we will see that this is the solution one gets when derives the equation as the hydrodynamical limit of the simple exclusion process.

3. The Simple Exclusion Process

The state space of the process is $\mathbf{X} := \{0,1\}^{\mathbb{Z}}$. Elements of $\mathbf{X}$ are functions that associate a number 0 or 1 to each integer. We call these elements configurations, denote them by greek letters η, ξ, σ, etc. and say, for a configuration η, that a site x is occupied by a particle if $\eta(x) = 1$ otherwise we say that x is empty. We identify a configuration η with the subset of $\mathbb{Z}$ $\{x : \eta(x) = 1\}$ of occupied sites. Measures on $\mathbf{X}$ are characterized on cylindric functions, *i.e.* functions that depend on a finite number of sites.

The process was described in the introduction. We proceed now to construct it using the graphical representation. This construction has been widely used in other particle systems such as the contact process, the voter model, the symmetric simple exclusion process and the branching exclusion process. In all those processes the main tool was duality. Here it is coupling.

At each bond $(x, x+1)$ associate a Poisson point process (Ppp) with rate 1. Each of these processes is a sequence of times, that we call $\omega(x,n)$, $n \in \mathbb{N}$ with the property that $\{\omega(x,n)-\omega(x,n-1)\}_{n,x}$ is a family of mutually independent random variables with exponential distribution with parameter 1. That is, $P(\omega(x,n) - \omega(x,n-1) > t) = e^{-t}$. Neglect the set of probability zero where at least two of these times coincide, *i.e.* the set $\{\omega :$ there exist $(x,n),(x',n')$ such that $\omega(x,n) =$

$\omega(x', n')\}$. We say that an arrow going from x to $x+1$ is present at the times $\omega(x,n)$, $n \in \mathbb{N}$. Call ω a configuration of arrows and $(\Omega, \mathcal{F}, P)$ the probability space induced by the Ppp described above.

Fix now a time $\bar{t}$. The set $\{\omega : \omega(x,1) < \bar{t}, \text{ for all } x > 0\}$ has probability zero, as well as the event defined in the same way but with $x < 0$. This means that for almost all ω there is a pair of sites $x, x+1$ such that there are no arrows connecting them in the interval $(0, \bar{t})$. Repeating the same argument, we can say that with probability one there is a sequence of sites x_i, $i \in \mathbb{Z}$ such that there are no arrows connecting x_i and x_i+1 in the time interval $(0, \bar{t})$. We consider only the ω belonging to this set of probability one. For each ω, we construct the process separately in the boxes $[x_i+1, x_{i+1}] \cap \mathbb{Z}$. Of course the x_i are function of ω and $\bar{t}$.

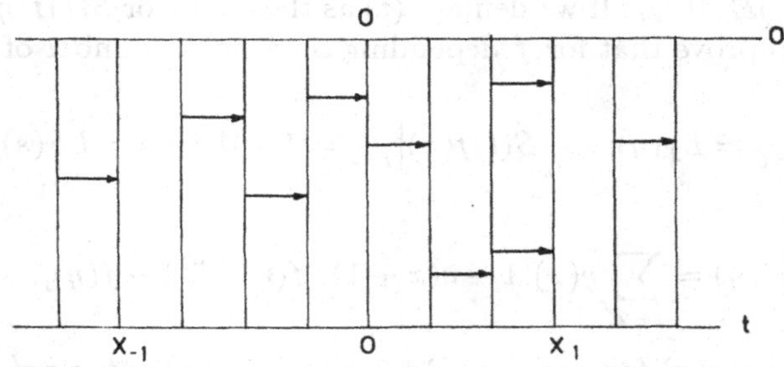

Figure 3. A typical realization of a ω with x_{-1} and x_1.

Now we put an initial configuration η at time zero, at the top of our graph –the time is flowing down– and construct η_t, $0 \le t \le \bar{t}$, as a function of η and ω. Since there are no arrows connecting different boxes, there is no interaction among them. Since the boxes have finite length, we are able to label the arrows inside each box by order of appearance. If the first arrow goes from, say, x to $x+1$, and before the arrow there is a particle at x and no particle at $x+1$, then the particle jumps from x to $x+1$ so that after the arrow there is a particle at $x+1$ and no particle at x. If before the arrow from x to $x+1$ there is a different event (two particles, two holes or a particle at $x+1$ and no particle at x), then nothing happens: the configuration after the arrow is exactly the same as before. Then we go to the second arrow and repeat the procedure up to the last arrow in the box and go to the next box constructing in this way $\eta^{\eta}_{\omega,t}$, $0 \le t \le \bar{t}$, with initial configuration η. For times greater than $\bar{t}$, use $\eta_{\bar{t}}$ as initial configuration and repeat the procedure to construct the process between $\bar{t}$ and $2\bar{t}$, and so on.

In Figure 4 we have drawn the trajectories of two particles starting at sites 0 and 1. They interact when the particle at 0 can not use the arrow as 1 is occupied. We denote by $E_\eta f(\eta_t)$ or $Ef(\eta^{\eta}_t)$ the expected value of $f(\eta_t)$ with respect to P when the initial configuration is η. If ν is a measure on $\mathbf{X}$, we denote

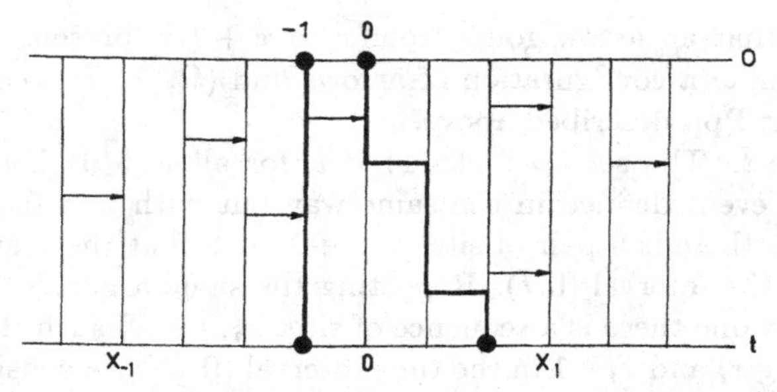

Figure 4. For the same ω, a realization of η_t.

$E_\nu f(\eta_t) := \int d\nu(\eta) E_\eta f(\eta_t)$. If we define $S(t)$ as the operator $S(t)f(\eta) := E_\eta f(\eta_t)$, it is an exercise to prove that for f depending on a finite number of coordinates,

$$\frac{d}{dt}S(t)f(\eta)\Big|_{t=0} = Lf(\eta); \quad \frac{d}{dt}S(t)f(\eta)\Big|_{t=s} = S(s)Lf(\eta) = LS(s)f(\eta) \tag{3.1}$$

where

$$Lf(\eta) = \sum_{x\in\mathbb{Z}} \eta(x)(1-\eta(x+1))[f(\eta^{x,x+1}) - f(\eta)] \tag{3.2}$$

where $\eta^{x,y}(z)$ equals $\eta(x)$ for $z = y$, $\eta(y)$ for $z = x$ and $\eta(z)$ for $z \neq x, y$.

The operator L is called the generator of the process and plays the role that probability transition functions do in discrete time Markov processes. It describes the intuitive instantaneous behavior of the process. In the Liggett existence theorem for interacting particle systems the process is constructed starting from the generator and proving the existence of a semigroup $S(t)$ satisfying (3.1), from where $E_\eta f(\eta_t)$ is defined. The existence theorem of Hille Yoshida is used. The graphical approach goes back to Harris [h2] and Bertein and Galves [bg], see Liggett [L].

4. The Hydrodynamical Limit

We describe here the heuristic derivation of equations (2.1) from the process η_t and state the theorems of convergence. We use the following notations: for a measure on $\mathbf{X}$, $\nu f := \int d\nu(\eta) f(\eta)$ and $\nu S(t)$ is the measure defined by $\nu S(t) f = \int d\nu(\eta) S(t) f(\eta)$. The product measures ν_α are the measures defined by

$$\nu_\alpha f_A = \alpha^{|A|}$$

where $f_A(\eta) := \prod_{x\in A} \eta(x)$. This means that a configuration η picked from the distribution ν_α can be constructed in the following way: at each site of $\mathbb{Z}$ put a particle with probability α. Do this independently for each site. Define $\eta_t^\varepsilon(r) :=$

$\eta_{\varepsilon^{-1}t}(\varepsilon^{-1}r)$, where $\varepsilon^{-1}r$ is an abuse of notation for integer part of $\varepsilon^{-1}r$. Applying (3.2) we get

$$\frac{d}{dt}E_\nu(\eta_t^\varepsilon(r)) = \varepsilon^{-1}E_\nu\left[-\eta_t^\varepsilon(r)(1-\eta_t^\varepsilon(r+\varepsilon)) + \eta_t^\varepsilon(r-\varepsilon)(1-\eta_t^\varepsilon(r))\right]$$

now, if there exist a limit $u(r,t) = \lim_{\varepsilon\to 0} E_\nu(\eta_t^\varepsilon(r))$ and the measure at time $\varepsilon^{-1}t$ is approximately a product measure, such that $E_\nu(\eta_t^\varepsilon(r)(1-\eta_t^\varepsilon(r+\varepsilon))$ converges, as $\varepsilon \to 0$, to $u(r,t)(1-u(r,t))$, then this limit must satisfy the Burgers equation (2.1).

In fact it is proven that if $u(r,t)$ is a solution of the Burgers equation (2.1) with $u(r,t) = u_0(r)$ and ν^ε is a family of product measures such that $\nu^\varepsilon(\eta(\varepsilon^{-1}r)) = u_0(r)$, then at the continuity points of $u(r,t)$,

$$\lim_{\varepsilon\to 0} \nu^\varepsilon S(\varepsilon^{-1}t)\tau_{\varepsilon^{-1}r} f = \nu_{u(r,t)} f \tag{4.1}$$

where $\tau_x\eta$ is the configuration defined by $\tau_x\eta(z) = \eta(z+x)$ and $\tau_x f(\eta) = f(\tau_x\eta)$. Eq. (4.1) has been proven first for initial profiles $u_0(r)$ that have only one discontinuity by Rost [r] in the non increasing case and extended by Benassi and Fouque [bf1] and Andjel and Vares [av] to the non decreasing case and for more general jump probabilities. Benassi, Fouque, Vares and Saada [bfvs] extended the result to any monotone profile and Landin [la2] to piecewise constant initial profiles presenting two discontinuities. We prove (4.1) when u_0 is a one step nondecreasing shock, like in (2.1) using the existence of a microscopic interface.

Related to that, the convergence of the density fields has been proven. Let Φ be a continuous function with compact support. Then,

$$\lim_{\varepsilon\to 0} \varepsilon \sum_{x\in\mathbb{Z}} \Phi(\varepsilon x)\eta_{\varepsilon^{-1}t}(x) = \int_{\mathbb{R}} \Phi(r)u(r,t)dr \tag{4.2}$$

P almost surely.

Notice that (4.1) gives the weak convergence of the process on the points of continuity of $u(r,t)$. Nothing is said about the points of discontinuity i.e. in $r = vt$. The expected result is the following

$$\nu_{\rho,\lambda}S(\varepsilon^{-1}t)\tau_{\varepsilon^{-1}v} = \tfrac{1}{2}\nu_\rho + \tfrac{1}{2}\nu_\lambda$$

This has been proven by Wick [w] in the case $\rho = 0$ for a related model that is isomorphic to this one and by De Masi, Kipnis, Presutti and Saada [dkps]. We give Wick's proof in Section 7. Then, using symmetry arguments, Andjel, Bramson and Liggett [abl] showed the result for the case $\lambda + \rho = 1$ *i.e.* $v = 0$. Presumably the study of the fluctuations of the microscopic shock allows one to treat this case also (see Section 13).

5. Invariant Measures

A measure ν is (time) invariant for the process if $\nu S(t) = \nu$ for all t. This means that if one starts the process with the initial measure ν and looks at the distribution at later times, one finds that the process is still distributed according to ν. Liggett [1] described the set of all invariant measures. This is a compact and convex set and is described by the extremal invariant measures. The product measures ν_α defined in the previous section are invariant for the process. To prove that, one has to show

$$\nu_\alpha S(t) f_A = \nu_\alpha f_A$$

or equivalently

$$\frac{d}{dt} \nu_\alpha S(t) f_A \big|_{t=0} = 0$$

Suppose that A is a block of sites, $A = [a, b] \cap \mathbb{Z}$, $a, b \in \mathbb{Z}$. One has to prove that the rate of entering $\mathbf{A} := \{\eta : \eta(x) = 1, x \in A\}$ is the same as the rate of exiting. But this is true because the rate of entering is $\nu_\alpha\{\eta(a-1) = 1, \eta(a) = 0, \eta(a+1) = 1, \ldots, \eta(b) = 1\} = \alpha^{b-a}(1-\alpha)$ and the rate of exiting is $\nu_\alpha\{\eta(a) = 1, \ldots, \eta(b) = 1, \eta(b+1) = 0\} = \alpha^{b-a}(1-\alpha)$. The same argument can be applied when A consists in more than one block.

There are also other measures, called "blocking" measures. These are also product but not translation invariant. This means that the density depends on the site. We call them $\nu^{(n)}$. They give mass one to a single configuration:

$$\nu^{(n)}(\eta^{(n)}) = 1$$

where

$$\eta^{(n)}(x) := \begin{cases} 1, & \text{if } x \geq n \\ 0, & \text{if } x < n \end{cases}$$

The $\nu^{(n)}$ are translation of each other.

Theorem 5.1. (Liggett). Every invariant measure for the process is a convex combination of ν_α, $0 \leq \alpha \leq 1$ and $\nu^{(n)}$, $n \in \mathbb{Z}$.

This proof is based on coupling and we will sketch it after a discussion of coupling in the next section.

It is easy to verify that the set of invariant measures is convex and compact in the topology of weak convergence. Hence all invariant measure is a convex combination of the extremal points of the set, by the Krein Millman Theorem. A measure is said to be invariant extremal if it can not be written as a non trivial convex combination of other invariant measures. We will see later that the extremal invariant measures play a role in the proof of law of large numbers.

6. Coupling

Coupling is the main tool in this process. Almost all the results have been proved using coupling. There is a simple way of describing it using the graphical representation.

Let η and η' be two initial configurations. The coupled process is the process defined by $(\eta_{\omega,t}, \eta'_{\omega,t})$. In other words we are realizing the two processes with the same realization of the arrows. This implies that the coupled process is realized in the same probability space $(\Omega, \mathcal{F}, P)$, and we can refer to probabilities or expectations with the same letters P and E. Since the particles of each copy follow the arrows without influence of the other copy, each marginal of the coupling has the distribution of the simple exclusion process. Intuitively the coupling works in the following way. Particles at site x of the configurations η_t and η'_t use the same arrows, so that they try to jump at the same time. If the destination site $x + 1$ is empty in both configurations the jump is realized in both marginals, but if it is occupied for one of the marginals, then the jump is realized only for the other marginal.

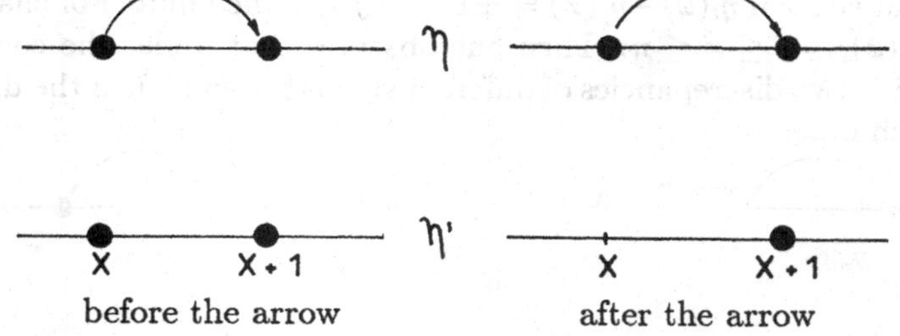

before the arrow after the arrow

Figure 5. Coupling.

In Figure 5 we see the configurations of η and η' before and after an arrow. One of the features of the coupling is that if $\eta \geq \eta'$ (coordinatewise) then $\eta_{\omega,t} \geq \eta'_{\omega,t}$ for almost all $\omega \in \Omega$, all $t \geq 0$. To prove this, it is sufficient to show that this is true for one arrow, as there is a finite number of arrows in each of the boxes where we construct the process. Suppose then that there is an arrow from x to $x + 1$. There are the following possibilities, the top line represents the occupation numbers of the configuration η_t at sites x and $x + 1$, before the arrow to the left and after the arrow to the right. The bottom line represents the same for the configuration η'. We ilustrate only the cases when $\eta \geq \eta'$:

11 11 11 11 11 11 11 11 10 01 10 01 01 01 01 01 00 00
10 01 01 01 11 11 00 00 10 01 00 00 01 01 00 00 00 00

We see that in any case, if before the jump the top configuration is greater than or equal to the bottom configuration, then the same is true after the jump. This property is called attractivity. This implies that

$$\nu \leq \mu \text{ implies } \nu S(t) \leq \mu S(t)$$

where $\nu \leq \mu$ is defined by the following equivalent statements

1. There exists $\tilde{\nu}$ on $\mathbf{X}^2$ with marginals ν and μ such that $\tilde{\nu}\{(\eta, \eta') : \eta \leq \eta'\} = 1$.

2. For all non decreasing f, $\nu f \leq \mu f$.

If $\rho \leq \lambda$, then $\nu_\rho \leq \nu_\lambda$. To see that, we construct a measure on $\mathbf{X}^2$ with marginals ν_ρ and ν_λ and concentrated on $\{(\eta, \eta') : \eta \leq \eta'\}$. We need a family $\{U_x\}_{x \in \mathbb{Z}}$ of random variables which are independent and uniformly distributed on $[0,1]$. Define $\eta(x) = 1\{U_x \leq \rho\}$ and $\eta'(x) = 1\{U_x \leq \lambda\}$. The resulting distribution of (η, η') has the desired properties.

Now we sketch how Liggett uses the coupling to describe the set of invariant measures for the process in the translation invariant case. First prove that if ν and ν' are extremal invariant for $S(t)$ then there exists an extremal invariant measure $\tilde{\nu}$ for the coupled process with marginals ν and ν'. We say that there is a discrepancy at site x if $\eta_t(x) - \eta'_t(x) = \pm 1$. Let f_n be the number of changes of sign of $\eta_t(x) - \eta'_t(x)$, $-n \leq x \leq n$. Then one observes that under the coupling, if an arrow involving two discrepancies of different sign is present, then the discrepancies eliminate each other.

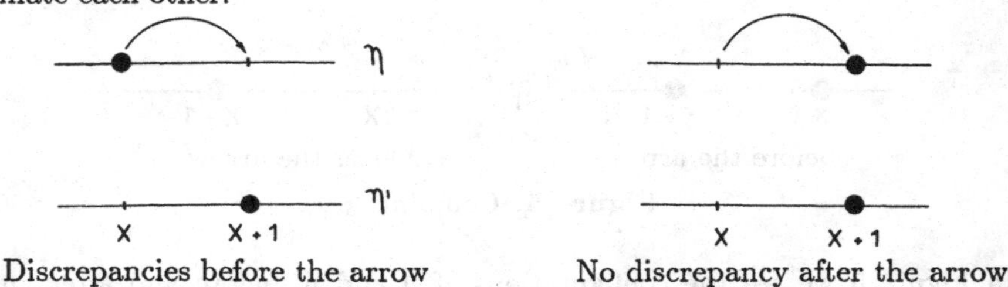

Discrepancies before the arrow   No discrepancy after the arrow

Figure 6. Discrepancies eliminate each other.

This is the key ingredient to prove that if $\tilde{\nu}$ is extremal invariant and translation invariant for the coupled process, then

$$\tilde{\nu}\{\eta \geq \eta'\} = 1 \quad or \quad \tilde{\nu}\{\eta \leq \eta'\} = 1. \tag{6.1}$$

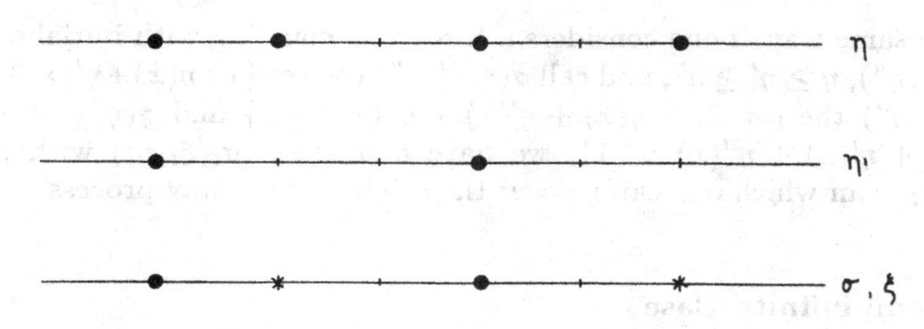

•: are σ-particles (first class)
*: are ξ-particles (second class)

Figure 7. Another way of looking at the coupling.

This implies that all extremal invariant and translation invariant measures are the ν_α's.

Later it will be useful to read the coupling in the following different way. Assume that (η, η') is the initial pair of configurations for the coupled process and that $\eta' \leq \eta$. Call $\xi(\eta, \eta')$ the set of sites $\{x : \eta(x) > \eta'(x)\}$ and $\sigma(\eta, \eta')$ the set $\{x : \eta(x) = \eta'(x)\}$.

Also call $\xi_t := \xi(\eta_t, \eta_t')$ and $\sigma_t := \sigma(\eta_t, \eta_t')$. Then we can recover (η_t, η_t') from (σ_t, ξ_t). On the other hand this last coupled process has an interesting property. First, at each site there is at most one particle, either ξ or σ. Second, when an arrow appears involving two sites with the same kind of particles (ξ-ξ or σ-σ), the same rules as for the (non coupled) process η_t hold. But when the arrow involves σ-ξ particles the rules are different. If the arrow goes from the σ particle to the ξ particle, then the particles interchange positions. If the arrow goes from the ξ to the σ particle, then nothing happens. In some way we can say that the σ particles "do not see" the ξ particles that must live in the sites not occupied by the σ particles. For this reason it is said that the σ particles are "first class particles" and the ξ particles are "second class", or that the σ particles have "priority" over the ξ particles. This priority is denoted $\sigma_t \vdash \xi_t$. This process can be constructed directly, just fixing the initial configurations σ and ξ and the priorities $\sigma_t \vdash \xi_t$.

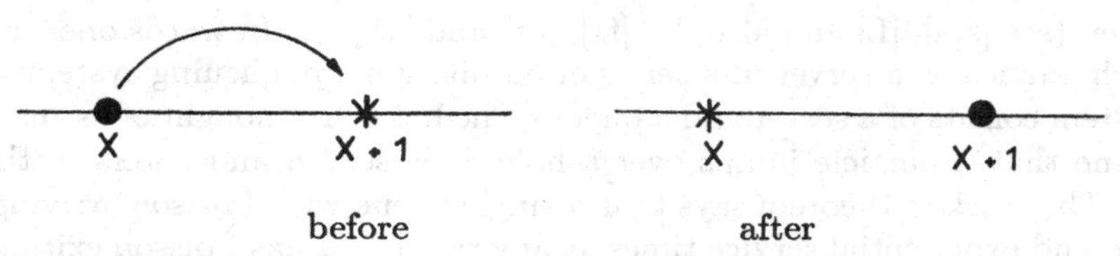

Figure 8. The σ - ξ version of Figure 5.

In the same way, if one considers a three way coupling with initial configurations (η,η',η''), $\eta \geq \eta' \geq \eta''$, and call $\sigma(\eta,\eta',\eta'')$ the set $\{x : \eta(x)+\eta'(x)+\eta''(x) = 3\}$, $\xi(\eta,\eta',\eta'')$ the set $\{x : \eta(x)+\eta'(x)+\eta''(x) = 2\}$ and $\gamma(\eta,\eta',\eta'')$ the set $\{x : \eta(x)+\eta'(x)+\eta''(x) = 1\}$, we have a process $(\sigma_t,\xi_t,\gamma_t)$ with priorities $\sigma_t \vdash \xi_t \vdash \gamma_t$ from which one can recover the original three way process.

7. The Semi-infinite Case

In this section we consider the measure $\nu_{0,\lambda}$, that is, the product measure with densities 0 and λ to the left and right of the origin, respectively. For convenience we assume also that there is a particle at the origin, and define $\nu' := \nu(.|\eta(0) = 1)$. Hence our initial measure is $\nu'_{0,\lambda}$. Call $X(t)$ the position at time t of the particle that is initially at the origin. We keep track of the position of that particle considering the process $(\eta_t, X(t))$ in $\{(\eta,z) : \eta \in \mathbf{X},\ z \in \mathbb{Z},\ \eta(z) = 1\}$. The generator of this process is the following

$$\bar{L}f(\eta,z) = \sum_{x,x+1\neq z} \eta(x)(1-\eta(x+1))[f(\eta^{x,x+1},z) - f(\eta,z)]$$
$$+ (1-\eta(z+1))[f(\eta^{z,z+1},z+1) - f(\eta,z)]$$

Now we are interested in the process as seen from the tagged particle. Hence we consider the process $\eta'_t := \tau_{X(t)}\eta_t$. This process has the following generator

$$L'f(\eta) = \sum_{x,x+1\neq 0} \eta(x)(1-\eta(x+1))[f(\eta^{x,x+1}) - f(\eta)]$$
$$+ (1-\eta(1))[f(\tau_1\eta^{0,1}) - f(\eta)]$$

For this process there is a translation each time that the tagged particle moves, in such a way that the tagged particle is always at the origin. The position of the tagged particle can be recovered from this process by defining $X(t) :=$ number of translations of the system in the time interval $[0,t]$. The following remarkable result is the key tool in this section. It comes from queuing theory. The connection between the simple exclusion process and a system of queues has been established by Kesten (see [spi], [L] and also [k], [f1], [df] and [dkps]). In words one can think that each particle is a server in a series of infinitely many queuing systems, where each system consits of a server and a queue. The holes are thought of as customers. Each time that a particle jumps over a hole, it is served and enters in the next system. The Burke's theorem says that a single queue with Poisson arriving times at rate a and exponential service times at any rate $b > a$ has Poisson exiting times at the same rate a. In this way the arriving time of the next system is the same as the one for the previous one and the exiting time of any system is Poisson. In our context, we have:

Theorem 7.1. Burke's Theorem. If the initial measure is $\nu'_{0,\lambda}$ then $\tau_{X(t)}\eta_t$ and $X(t)$ are independent. Indeed

$$E(f(\tau_{X(t)}\eta_t)|X(t)) = \nu'_{0,\lambda}f \text{ for all } t \geq 0$$

and $X(t)$ is a nearest neighbor totally asymmetric random walk with parameter $1-\lambda$.

Proof. The measure $\nu'_{0,\lambda}$ is invariant for the process $\eta'_t := \tau_{X(t)}\eta_t$. Define the reversed process η^*_s, $0 \leq s \leq t$, with respect to $\nu'_{0,\lambda}$, by

$$E_{\nu'_{0,\lambda}} f(\eta^*_s) := \int d\nu'_{0,\lambda}(\eta) E(f(\eta'_{t-s})|\eta'_t = \eta)$$

Since the measure $\nu'_{0,\lambda}$ is stationary, η^*_s is a stationary Markov process with that measure as invariant. The generator L^* is the adjoint of the operator L' on $\mathbf{L}_2(\nu'_{0,\lambda})$ and is given by

$$\begin{aligned} L^* f(\eta) = &\sum_{x,x-1\neq 0} \eta(x)(1-\eta(x-1))[f(\eta^{x,x-1}) - f(\eta)] \\ &+ (1-\lambda)[f(\tau_{-1}\eta^{0,-1}) - f(\eta)] \end{aligned}$$

To prove that L^* is the adjoint of L' in $\mathbf{L}_2(\nu'_{0,\lambda})$ one has to prove that $\nu'_{0,\lambda}(fLg) = \nu'_{0,\lambda}(gL^*f)$. This follows from the following identities:

$$\begin{aligned} &\int \nu'_{0,\lambda}(d\eta)(\eta(x)(1-\eta(x+1))f(\eta)g(\eta^{x,x+1})) \\ &= \int \nu'_{0,\lambda}(d\eta)(\eta(x+1)(1-\eta(x))f(\eta^{x,x+1})g(\eta)) \end{aligned}$$

for $x, x+1 \neq 0$, and

$$\int \nu'_{0,\lambda}(d\eta)((1-\eta(1))f(\eta)g(\tau_1\eta^{0,1})) = \int \nu'_{0,\lambda}(d\eta)(1-\lambda)f(\tau_{-1}\eta^{0,-1})g(\eta))$$

Now observe that out of the origin η^*_t is a simple exclusion process with jumps only from right to left. Furthermore the leftmost particle jumps to the left not with rate one but with rate $1-\lambda$, producing a translation of the system so that we keep the configuration in the set with one particle at the origin and no particles to its left. Call $X^*(t)$ the number of translations of η^*_t in the interval $[0,t]$. Then $(\eta'_t, X(t)-X(0)) = (\eta^*_0, X^*(0)-X^*(t))$ in distribution. Now, since the law of $-X^*(t)$ is that of a Ppp of parameter $1-\lambda$ and is independent of η^*_0, so is the law of $X(t)$. Since $\eta^*_0 = \eta_t$, this finishes the proof. ♣

The next result is just a consequence of the fact that $X(t)$ is a Poisson point process and can be found in introductory books on stochastic processes.

Corollary 7.2. Law of large numbers and Central limit theorem. The following hold

$$\lim_{t\to\infty} \frac{X(t)}{t} = 1-\lambda, \; P_{\nu_{0,\lambda}} \; a.s.$$

$$\lim_{\varepsilon\to 0} \frac{X(\varepsilon^{-1}t)-(1-\lambda)\varepsilon^{-1}t}{\sqrt{(1-\lambda)\varepsilon^{-1}t}} = \mathcal{N}(0,1)$$

Corollary 7.3. Hydrodynamics. The following holds

$$\lim_{\varepsilon\to 0} \nu_{0,\lambda}S(\varepsilon^{-1}t)\tau_{\varepsilon^{-1}r} = \begin{cases} \nu_0 & \text{if } r<1-\lambda \\ \nu_\lambda & \text{if } r>1-\lambda \end{cases}$$

Proof. Write

$$\tau_{\varepsilon^{-1}r}\nu'_{0,\lambda}S(\varepsilon^{-1}t)f = E_{\nu'_{0,\lambda}}\tau_{\varepsilon^{-1}r-X(\varepsilon^{-1}t)}f(\tau_{X(\varepsilon^{-1}t)}\eta_{\varepsilon^{-1}t})$$

Now, by Burke's theorem, $\tau_{X(t)}\eta_t$ has distribution $\nu'_{0,\lambda}$ for all t independently of $X(t)$. On the other hand, by the law of large numbers for $X(t)$, $\varepsilon^{-1}r - X(\varepsilon^{-1}t)$ converges almost surely, as $\varepsilon\to 0$, to $-\infty$ if $r>vt$ and to ∞ if $r<vt$. This proves the Corollary. ♣

Corollary 7.4. Dynamical phase transition. The following holds

$$\lim_{\varepsilon\to 0} \nu_{0,\lambda}S(\varepsilon^{-1}t)\tau_{\varepsilon^{-1}(1-\lambda)t+\varepsilon^{-1/2}r} = (1-g(r))\nu_0 + g(r)\nu_\lambda$$

where $g(r) := (2\pi)^{-1/2}\int_{-\infty}^{r}\exp(-x^2/2)dx$ is the standard Gaussian distribution function.

Proof. The idea is the same as in the proof of the previous corollary. The difference is that in order to show that $|X(\varepsilon^{-1}t)-\varepsilon^{-1}vt-\varepsilon^{-1/2}r|$ diverges one needs to use the fact that $X(t)$ is roughly a Gaussian random variable with variance (of the order of) t. After this it suffices to notice that $g(r)$ is just the probability that $X(\varepsilon^{-1}t)$ is to the left of $\varepsilon^{-1}vt + \varepsilon^{-1/2}r$ and apply Burke's theorem again. ♣

8. The Tagged Particle Process

This is the process $(\eta_t, X(t))$ on the state space $\mathbf{X} \times \mathbb{Z}$. If $X(0) = x$ then $\eta_0(x) = 1$, so that $X(t)$ describes the position of the particle that at time 0 was at x. The generator for this process is

$$\bar{L}f(\eta, z) = \sum_{x,x+1\neq z} \eta(x)(1-\eta(x+1))[f(\eta^{x,x+1}, z) - f(\eta, z)]$$
$$+ (1-\eta(z+1))[f(\eta^{z,z+1}, z+1) - f(\eta, z)]$$

In this process $\eta_t(X(t)) = 1$ for all t. Notice that by the translation invariance of P (i.e. $\tau_x\omega$ has the same probability as ω), we get

$$E_{(\eta,x)}\tau_u f(\eta_t, X(t)) = E_{\tau_u(\eta,x)} f(\eta_t, X(t)) \tag{8.1}$$

where $\tau_u(\eta, x) := (\tau_u\eta, x-u)$. Now define the process as seen from the tagged particle

$$\eta'_t = \tau_{X(t)}\eta_t$$

For this process $\eta'_t(0) = 1$ for all t. The generator is

$$L'f(\eta) = \sum_{x,x+1\neq 0} \eta(x)(1-\eta(x+1))[f(\eta^{x,x+1}) - f(\eta)]$$
$$+ (1-\eta(1))[f(\tau_1\eta^{0,1}) - f(\eta)]$$

Let $S(t)$ be the semigroup for η_t and $S'(t)$ the semigroup for η'_t. Let $\mathcal{S}$ be the set of translation invariant measures, *i.e.* $\mathcal{S} := \{\mu \text{ on } \mathbf{X} : \mu = \tau_x\mu \text{ for all } x\}$. Let $\mu \in \mathcal{S}$ with $\mu(\eta(0)) > 0$. Define $\mu' := \mu(.|\eta(0) = 1)$. That is, for any cylindric f,

$$\mu' f = \frac{1}{\mu(\eta(0))} \int d\mu(\eta)\eta(0)f(\eta).$$

The next result is due to Harris [h1] (see also Port Stone [ps] and Ferrari [f1]). It says that the distribution of the process as seen from the tagged particle is the same as the distribution of the process as seen from the origin conditioned to have a particle at the origin.

Theorem 8.2. Let μ be translation invariant and $\mu(\eta(0)) > 0$, then $\mu' S'(t) = (\mu S(t))'$.

Proof. First show, as an exercice, that $\mu S(t)(\eta(0)) = \mu(\eta(0))$. For this it suffices to show that $\mu L(\eta(0)) = 0$. By the definition of $(.)'$,

$$\begin{aligned}
(\mu S(t))' f &= \frac{1}{\mu S(t)(\eta(0))} \int d(\mu S(t))(\eta) f(\eta)\eta(0) \\
&= \frac{1}{\mu(\eta(0))} \int d\mu(\eta) \sum_{x \in \mathbb{Z}} \eta(x) E_{(\eta,x)}[f(\eta_t) 1\{X(t) = 0\}] \\
&= \frac{1}{\mu(\eta(0))} \int d\mu(\eta) \sum_{x \in \mathbb{Z}} \eta(x) E_{(\tau_x \eta,0)}[f(\tau_{-x}\eta_t) 1\{X(t) = -x\}] \\
&= \frac{1}{\mu(\eta(0))} \sum_{x \in \mathbb{Z}} \int d\mu(\eta)\eta(x) E_{(\tau_x \eta,0)}[f(\tau_{-x}\eta_t) 1\{X(t) = -x\}]
\end{aligned} \tag{8.3}$$

where the third identity follows from the translation invariance of the process (8.1). Now writing $\eta(x) = (\tau_x \eta)(0)$ and using the translation invariance of μ, the last member of (8.3) equals

$$= \frac{1}{\mu(\eta(0))} \int d\mu(\eta)\eta(0) E_{(\eta,0)} \sum_{x \in \mathbb{Z}} f(\tau_{-x}\eta_t) 1\{X(t) = -x\} = \mu' S'(t) f$$

Note: all interchanges of sums with integrals are justified as f is cylindric (hence bounded). ♣

Corollary 8.4. If μ is invariant for $S(t)$ and translation invariant, then μ' is invariant for $S'(t)$. In particular we have that ν'_ρ is invariant for the process as seen from the tagged particle.

Proof. $\mu' S'(t) = (\mu S(t))' = \mu'$. ♣

Remark. Identify a configuration η with the ordered set $\{x_i\}_{i \in \mathbb{Z}}$ of occupied sites such that x_0 is the first occupied site to the right of the origin of η. It has been proven [f1] that $\{\nu'_\rho : 0 \leq \rho \leq 1\}$ is the set of extremal invariant and "translation invariant" measures –the measures for which η and $\tau_{x_i}\eta$ are identically distributed.

There are also other invariant measures that concentrate on the set $\{\eta : \sum_{x<0} \eta(x) < \infty\}$. These can be constructed in the following way: let ξ be a configuration distributed according to ν'_λ. Now call x_i the position of the i-th ξ

particle ($x_0 = 0$). Let $\eta = \eta(\xi)$ be defined by $\eta(x) = 1\{x \geq 0\}\xi(x)$. Then the distribution of η is $\nu'_{0,\lambda}$. Call $\mu^{[i]}$ the distribution of $\tau_{x_i}\eta$. It is proven in [fl] that all invariant measures for the process as seen from the tagged particle are convex combinations of ν'_α and $\mu^{[i]}$, but we don't use it in this paper.

Now, calling $G(t)$ the position of the leftmost particle of η_t so constructed and $X(t)$ the position of the tagged particle of ξ we get that $G(t) \equiv X(t)$. Hence, if the law of large numbers or the central limit theorem holds for one of these positions then it also holds for the other.

9. Laws of Large Numbers

To prove laws of large numbers we use the ergodic theorem. We follow Kipnis [k] and Saada [s1] [s2]. The Birkhoff ergodic theorem guarantees that if μ is invariant, then for all cylindric f, $(1/t)\int_0^t f(\eta_s^\eta)ds$ converges P_μ a.s. to a limit, where we used the notation η_s^η to indicate the random configuration obtained at time s when the initial configuration is η.

We say that P_μ is ergodic if for all cylindric f,

$$\hat{f}(\eta) := \lim_{t\to\infty} t^{-1} \int_0^t f(\eta_s^\eta)ds = \mu f \quad P_\mu \ a.s.$$

Lemma 9.1. If μ is extremal invariant for η_t then P_μ is ergodic.

Proof. We follow Rosenblatt [rb]. Assume that P_μ is not ergodic. Then there exists $c > 0$ and a cylindric function f such that $\mu(\mathbf{A}) = \beta$, $0 < \beta < 1$, where $\mathbf{A} := \{\eta : \hat{f}(\eta) > c\}$. But $\mathbf{A}$ is a.s. closed for the motion, *i.e.* for all $\eta \in \mathbf{A}$, $P_\eta(\eta_t \in \mathbf{A}) = 1$. The same is true for $\mathbf{A^c}$. Hence $\mu(.|\mathbf{A})$ and $\mu(.|\mathbf{A^c})$ are invariant and we can write $\mu = \beta\mu(.|\mathbf{A}) + (1-\beta)\mu(.|\mathbf{A^c})$. This implies that μ is not extremal. ♣

Theorem 9.2. Let $F(t)$ be an additive functional of η_t. That is, law$(F(t+s) - F(t)|\eta_u, u \leq t) = $ law$(F(t+s) - F(t)|\eta_t)$. Assume that μ is extremal invariant for the process and that $\lim_{h\to 0} h^{-1}E_\mu(F(t+h) - F(t))^2 < \infty$. Then

$$\lim_{t\to\infty} \frac{F(t)}{t} = \mu\psi \quad P_\mu \ a.s.$$

where ψ is the compensator of $F(t)$ defined by

$$\psi(\eta) := \lim_{h\to 0} \frac{E(F(t+h) - F(t)|\eta_t = \eta)}{h}$$

We say that $\psi(\eta)$ is the instantaneous increment of $F(t)$ when the configuration at time t is η.

Proof. Define

$$M(t) := F(t) - \int_0^t \psi(\eta_s)ds \tag{9.3}$$

It is easy to see that $M(t)$ is a martingale because the instantaneous increment of $F(t)$ is $\psi(\eta_t)$ which is exactly the same as the instantaneous increment of the integral in (9.3). Now, since $M(t)$ is a martingale, $t^{-1}E_\mu M(t)^2$ is independent of t and we can write

$$\frac{E_\mu M(t)^2}{t} = \lim_{h\to 0} \frac{E_\mu(M(t+h)-M(t))^2}{h} = \lim_{h\to 0} \frac{E_\mu(F(t+h)-F(t))^2}{h}$$

If this last limit is finite, we call it $\mu\Phi$ and we can apply the convergence theorem for martingales (Breiman [B]) to conclude that

$$\lim_{t\to\infty} \frac{M(t)}{t} = 0 \;\; P_\mu \;\; a.s.$$

So that

$$\lim_{t\to\infty} \frac{F(t)}{t} = \lim_{t\to\infty} t^{-1} \int_0^t \psi(\eta_s)ds = \mu\psi \;\; P_\mu \;\; a.s.$$

where the second identity follows from Lemma 9.1. ♣

Let's see some applications of this theorem.

Corollary 9.4. Law of large numbers for the flux. Let $F(t) := \#\{i : x_i(0) \le 0$ and $x_i(t) > 0\} - \#\{i : x_i(0) > 0$ and $x_i(t) \le 0\}$, where $x_i(t)$ is the position of the i-th particle of η_t. Then

$$\lim_{t\to\infty} \frac{F(t)}{t} = \lambda(1-\lambda) \;\; P_{\nu_\lambda} \;\; a.s.$$

Proof. We apply the theorem to the process η_t. We saw before that ν_λ is extremal invariant. On the other hand $\psi(\eta) = \eta(0)(1-\eta(1))$, $\nu_\lambda\psi = \lambda(1-\lambda)$ and $\nu_\lambda\Phi = \lambda(1-\lambda) < \infty$. ♣

The next corollary has been proven in Section 7, using Burke's theorem. We give a proof that does not use that result.

Corollary 9.5. Law of large numbers for the tagged particle.

$$\lim_{t\to\infty} \frac{X(t)}{t} = (1-\lambda) \quad P_{\nu'_\lambda} \ a.s.$$

Proof. We apply the theorem to the process $\tau_{X(t)}\eta_t$. It was proved by [f1] that ν'_λ is extremal for $\tau_{X(t)}\eta_t$. We have that $\psi(\eta) = (1-\eta(1))$, $\nu_\lambda\psi = (1-\lambda)$, $\lim_{h\to 0} E_{\nu_\lambda}(X(t+h)-X(t))^2/h = (1-\lambda)$. Hence $EM(t)^2 = t(1-\lambda) < \infty$ and the Theorem applies. ♣

Corollary 9.6. Law of large numbers for the flux throught a random position. Let $U(t)$ be a birth process with rate w independent of η_t , i.e. $P(U(t)=k) = e^{-wt}(wt)^k/k!$. Let $F(t)$ be the net flux throught $U(t)$: $F(t) := \#\{i : x_i(0) \le 0$ and $x_i(t) > U(t)\} - \#\{i : x_i(0) > 0$ and $x_i(t) \le U(t)\}$. Then

$$\lim_{t\to\infty} \frac{F(t)}{t} = \lambda(1-\lambda) - w \quad P_{\nu_\lambda} \ a.s.$$

Proof. We apply the theorem to the process $\tau_{U(t)}\eta_t$. First we claim that if μ is invariant for this process, then μ is translation invariant. It is easy to show that if μ is invariant then $\tau_x\mu$ is also invariant. To prove the claim it suffices to show that the processes with initial measure μ and $\tau_x\mu$ converge to the same measure. To do that we couple two versions of the process in the following way: we choose the same initial configuration for the two processes according to μ and consider two versions of $U(t)$: $U_0(t)$ with initial point 0 and $U_x(t)$ with initial point x. These two versions are independent up to $T :=$ first time that $U_0(t) = U_x(t)$. After T, they continue together. In that way $\tau_{U_0(t)}\eta_t$ and $\tau_{U_x(t)}\eta_t$ are the processes with initial measure μ and $\tau_x\mu$, respectively. Since $U_0(t)$ and $U_x(t)$ are independent random walks, they will meet with probability one and the two processes will coincide after T. Since the initial measures are invariant they must be the same. This proves the claim.

Now, using the method of Liggett described in Section 6, one proves that all extremal invariant measures for this process are the product measures with density ρ, $0 \le \rho \le 1$. Once proved that ν_λ is extremal invariant for this process, the rest follows as in the previous corollaries. ♣

10. Microscopic Shock in the General Case

We saw that in the case of vanishing left density the position of the leftmost particle defines a microscopic shock. Indeed the system as seen from the shock has a measure $\mu' S'(t) = \nu'_{0,\lambda}$ for all times. Such strong statment is not true in the case of non vanishing left density. We introduce a weaker definition of a microscopic shock:

Definition 10.1. We say that a random position $X(t)$ is a microscopic shock for η_t if calling μ_t the distribution of $\tau_{X(t)}\eta_t$, the following weak limits hold uniformly in t:

$$\lim_{x \to +\infty} \tau_x \mu_t = \nu_\lambda, \quad \lim_{x \to -\infty} \tau_x \mu_t = \nu_\rho$$

When the left density does not vanish it is not obvious how to define the position of a microscopic shock. One can try to tag a particle and follow it, but one immediately realizes that the tagged particle has the wrong velocity: in regions of ν_ρ has velocity $(1-\rho)$ and in regions of ν_λ has velocity $(1-\lambda)$. Nevertheless, in some sense, the idea of considering a last particle used for vanishing left density also works in the general case. We couple two processes. The σ process with initial measure ν_ρ and the $\eta = \sigma + \gamma$ process with initial measure $\nu_{\rho,\lambda}$. At time 0 we couple the initial configurations as in Section 5: Let $\{U_x\}_{x \in \mathbb{Z}}$ be a sequence of independent identically distributed random variables with distribution uniform in $[0,1]$. Given a realization of those variables we define $\sigma(x) = 1\{U(x) \le \rho\}$ and $\gamma(x) = 1\{\rho < U_x \le \lambda\}.1\{x \ge 0\}$. The two processes use the same arrows, hence the process σ_t coincides with the set of sites occupied by the two marginals and γ_t the set of sites occupied only by the marginal $\sigma + \gamma$. The reader can check that when an arrow appear from site x to site y, the following can happen:

1. The site x is occupied by a σ or γ particle and y is empty. Then the particle jumps.
2. The site x is occupied by a σ or γ particle and y is occupied by a σ particle. Then nothing happens.
3. The site x is occupied by a γ particle and y is occupied by a γ particle. Then nothing happens.
4. The site x is occupied by a σ particle and y is occupied by a γ particle. Then the particles interchange positions: after the arrow $\sigma(y) = 1$ and $\gamma(x) = 1$.

We say that the σ particles are first class and the γ are second class particles. We denote this priority by $\sigma_t \vdash \gamma_t$.

Our program is to prove that the law of the system as seen from the leftmost γ particle satisfies (10.1). When $\rho = 0$ the η process coincides with the γ process and we saw in Section 7 that this is the case. For $\rho > 0$ we call $X(t)$ the position of the leftmost γ particle. We have that $\sigma_t \vdash (\gamma_t \setminus X(t)) \vdash X(t)$ and since $\eta_t = \sigma_t + \gamma_t$, we have $(\eta_t \setminus X(t)) \vdash X(t)$, where we identify the position $X(t)$ with the configuration having a particle at $X(t)$ and no particles at the other sites. So the process $(\eta_t, X(t))$ is well defined without using the σ and γ processes.

Theorem 10.2. Assume that η_0 has distribution $\nu_{\rho,\lambda}$ and $X(0) = 0$. Then $X(t)$ is a shock for η_t.

The proof of this theorem uses properties of translation invariant systems. To get a related translation invariant system we couple the processes (σ_t, γ_t) with a third process ζ_t in such a way that $\sigma_t + \gamma_t + \zeta_t$ at time zero has distribution ν_λ that is translation invariant. The way of doing that is to use the uniform variables U_x: define $\zeta(x) = 1\{\rho < U_x \leq \lambda\}.1\{x < 0\}$. Call $\xi_t = \gamma_t + \zeta_t$. Then the process (σ_t, ξ_t) has initial product distribution with marginals ν_ρ and $\nu_{\lambda-\rho}$ and a translation invariant distribution for all t. We say that a measure π_2 on $\mathbf{X}^2$ has the good marginals if $\int d\pi_2(\sigma,\xi) f(\sigma) = \nu_\rho f$ and $\int d\pi_2(\sigma,\xi) f(\sigma+\xi) = \nu_\lambda f$. If the process (σ_t, ξ_t) has initial distribution with the good marginals, then at any time t the distribution of the process has the good marginals.

Assume now that $\xi(0) = 1$, so that $X(t)$ is the position of the ξ particle that at time zero is at the origin. Now we have the following result the proof of which is the same as the one of Theorem 8.2:

Theorem 10.3. Let $\lambda > \rho$, π_2 be a translation invariant measure, $\pi_2' = \pi_2(.|\xi(0) = 1)$ and $S_2'(t)$ be the semigroup of the process $\tau_{X(t)}(\sigma_t, \xi_t)$, then

$$\pi_2' S_2'(t) = (\pi_2 S_2(t))' \tag{10.4}$$

Equation (10.4) is crucial. It expresses the distribution of the process as seen from $X(t)$ in terms of the distribution of the process without shifting.

Lemma 10.5. If π_2 is a measure with the good marginals then for any cylindric function f on $\mathbf{X}$,

$$\begin{aligned} \lim_{x\to\pm\infty} \int d\pi_2'(\sigma,\xi)\tau_x f(\sigma) &= \nu_\rho f \\ \lim_{x\to\pm\infty} \int d\pi_2'(\sigma,\xi)\tau_x f(\sigma+\xi) &= \nu_\lambda f \end{aligned} \tag{10.6}$$

Proof. We use our knowledge that the σ and $\sigma + \xi$ marginals of π_2 are ν_ρ and ν_λ respectively. Call $g(x) = \tau_x f$. Since f is cylindric function, $g(x)$ is uniformly bounded and for any sequence x_n, there exists a subsequence $x_{n(k)}$ such that

$\pi_2' g(x_{n(k)})$ is convergent. We want to show that the limit is always $\nu_\rho f$. Since $\pi_2' g(x_{n(k)})$ is convergent, so is the Cesaro limit

$$\lim_{K\to\infty} (1/K) \sum_{k=1}^{K} \pi_2' g(x_{n(k)}) \tag{10.7}$$

To compute this limit we take yet another subsequence with the property that for all k, $n(k+1) - n(k) > b - a$, where we let $[a, b]$ be a (finite) interval containing the set of sites determining f. For this subsequence we have that, under π_2, $g(x_{n(k)})$ and $g(x_{n(k+1)})$ are independent. Then, since the σ marginal of π_2 is ν_ρ, the law of large numbers imply that

$$\lim_{K\to\infty} (1/K) \sum_{k=1}^{K} g(x_{n(k)}) = \nu_\rho f \quad \pi_2 \ a.s.$$

Since π_2' is absolutely continuous with respect to π_2, the same is true π_2' a.s. By dominated convergence we have that the limit in (10.7) must be $\nu_\rho f$ and we have proved the first line of (10.6). Analogously one proves the second line. ♣

Proof of Theorem 10.2. Let π_2 be the product measure with the good marginals. Let (σ, ξ) be distributed according to π_2'. Then, defining η by $\eta(x) = \sigma(x) + \xi(x)1\{x \geq 0\} = \sigma(x) + \gamma(x)$, we have that η has distribution $\nu'_{\rho,\lambda}$, $X(0) = 0$ and $\eta_t(x) = \sigma_t(x) + \xi_t(x)1\{x \geq X(t)\}$. Now, $\tau_{X(t)}\eta_t(x) = \tau_{X(t)}(\sigma_t(x) + \xi_t(x))1\{x \geq 0\}$. On the other hand by (10.4), $\tau_{X(t)}(\sigma_t, \xi_t)$ has distribution $(\pi_2 S_2(t))'$ which is absolutely continuous with respect to $\pi_2 S_2(t)$, a measure with the good marginals. Hence Lemma 10.5 implies that, uniformly in t,

$$\lim_{x\to+\infty} \int d(\nu'_{\rho,\lambda} S'(t))(\eta) \tau_x f(\eta) = \lim_{x\to+\infty} \int d(\pi_2' S_2'(t))(\sigma, \xi) \tau_x f(\sigma + \xi) = \nu_\lambda f$$

by Lemma 10.5. Analogously,

$$\lim_{x\to-\infty} \int d(\nu'_{\rho,\lambda} S'(t))(\eta) \tau_x f(\eta) = \lim_{x\to-\infty} \int d(\pi_2' S_2'(t))(\sigma, \xi) \tau_x f(\sigma) = \nu_\rho f \quad ♣$$

11. Law of Large Numbers for the Shock

In this section we prove that $X(t)/t$ converges almost surely to $v = 1 - \rho - \lambda$. Given $U(t) \in I\!R$ we define the σ flux throught $U(t)$ by $F_\sigma(t) := \#\{i : z_i(0) <$

$0, z_i(t) \geq U(t)\} - \#\{i : z_i(0) \geq 0, z_i(t) < U(t)\}$, where $z_i(t)$ is the position of the i-th particle of σ_t. For the uniqueness of the representation we fix $z_0(0)$ as the position of the first particle to the right of the origin. Analogously we define $F_{\sigma+\xi}(t)$ and $F_\xi(t)$. If π_2 is a measure with the good marginals, then σ_t and $\sigma_t + \xi_t$ are simple exclusion processes with (extremal invariant) measure ν_ρ and ν_λ respectively. Hence, by Corollary 9.6, if $U(t)$ is a Poisson point process of parameter w, then

$$\lim_{t\to\infty} \frac{F_{\sigma+\xi}(t)}{t} = \lambda(1-\lambda) - w\lambda$$

$$\lim_{t\to\infty} \frac{F_\sigma(t)}{t} = \rho(1-\rho) - w\rho$$

But $F_{\sigma+\xi}(t) = F_\sigma(t) + F_\xi(t)$, hence

$$\begin{aligned}\lim_{t\to\infty} \frac{F_\xi(t)}{t} &= [\lambda(1-\lambda) - \rho(1-\rho] - w(\lambda-\rho)\\ &= (\lambda-\rho)(v-w)\end{aligned}$$

where $v = 1 - \lambda - \rho$. The limit is negative for $w > v$ and positive for $w < v$. On the other hand, $F_{(.)}(t)$ is non increasing in $U(t)$ and for $U(t) = X(t)$, due to the exclusion interaction, $F_\xi(t) \equiv 0$. Hence

$$\lim_{t\to\infty} \frac{X(t)}{t} = v, \quad P_{\nu_{\rho,\lambda}}\ a.s. \clubsuit \tag{11.1}$$

As a consequence of the law of large numbers we can prove the hydrodynamics announced in Section 4:

Theorem 11.2. Let $u(r,t)$ be the following solution of the Burgers equation (2.1):

$$u(r,t) = \begin{cases} \lambda & \text{for } r > vt \\ \rho & \text{for } r < vt. \end{cases}$$

Then in the continuity points of $u(r,t)$,

$$\lim_{\varepsilon\to 0} \nu_{\rho,\lambda} S(\varepsilon^{-1}t)\tau_{\varepsilon^{-1}r} f = \nu_{u(r,t)} f$$

Proof. Let $A \subset \mathbb{Z}$ be a finite set and let $f_A(\eta) := \prod_{x\in A} \eta(x)$.

$$\begin{aligned}&\int d\nu_{\rho,\lambda}(\eta) E_\eta \tau_{\varepsilon^{-1}r} f_A(\eta_{\varepsilon^{-1}t})\\ &= \int d\nu_2'(\sigma,\xi) E_{(\sigma,\xi)} \tau_{\varepsilon^{-1}r} f_A(\sigma_{\varepsilon^{-1}t} + \gamma_{\varepsilon^{-1}t})\\ &= \int d\nu_2'(\sigma,\xi) E_{(\sigma,\xi)} \tau_{\varepsilon^{-1}r - X(\varepsilon^{-1}t)} f_A(\sigma'_{\varepsilon^{-1}t} + \gamma'_{\varepsilon^{-1}t})\\ &= \int d\nu_2'(\sigma,\xi) E_{(\sigma,\xi)} \sum_x \tau_{\varepsilon^{-1}r - x} f_A(\sigma'_{\varepsilon^{-1}t} + \gamma'_{\varepsilon^{-1}t}) 1\{X(\varepsilon^{-1}t) = x\}\end{aligned} \tag{11.3}$$

where E_η, is the expected value of the process with initial configuration η, etc. and $\gamma_t(z) = \xi_t(z)1\{z \geq X(t)\}$. Now, consider a positive number a such that $|r - vt| > a$, and decompose the sum in the last line of (11.3) in three parts: $\{x : |\varepsilon^{-1}r - x| \leq a\}$, $\{x : \varepsilon^{-1}r - x > a\}$ and $\{x : \varepsilon^{-1}r - x < -a\}$. The integral of the first of those three sums goes to zero by (11.1) and dominated convergence. By Lemma 10.5, the second part converges to $\nu_\lambda f_A$, if $r > vt$ and to zero otherwise and the third part converges to $\nu_\rho f_A$ if $r < vt$ and to zero otherwise. ♣

The following result can be proven in a similar way.

Theorem 11.4. Convergence of the density fields. For any smooth function Φ with compact support, the following holds

$$\lim_{\varepsilon\to 0} \varepsilon \sum_{x\in\mathbb{Z}} \Phi(\varepsilon x)\eta_{\varepsilon^{-1}t}(x) = \int_{\mathbb{R}} \Phi(r)u(r,t)dr$$

$P_{\nu_{\rho,\lambda}}$ almost surely.

12. Second Class Particles and Characteristics

We saw in Section 2 that in regions where the solution u of (2.1) is constant, the characteristics have speed $1 - 2u$, and that the shock forms at the meeting point of the characteristics with different speeds. In this section we study the microscopic counterpart of this phenomenon. We first show that if a second class particle is added at time zero at a given site, and the first class particles are distributed according to ν_α, then, calling $W(t)$ the position at time t of the second class particle, we have that

$$\lim_{t\to\infty} \frac{W(t)}{t} = 1 - 2\alpha, \quad P_{\hat{\nu}_\alpha}\ a.s. \tag{12.1}$$

Denote by $W(\varepsilon^{-1}r, \varepsilon^{-1}t)$ the position at time $\varepsilon^{-1}t$ of a second class particle that at time zero is at site (integer part of) $\varepsilon^{-1}r$. Then we prove that

$$\lim_{\varepsilon\to 0} \varepsilon W(\varepsilon^{-1}r, \varepsilon^{-1}t) = \begin{cases} w(r,t), & \text{for } t < t(r) \\ vt, & \text{for } t > t(r) \end{cases} \quad P_{\hat{\nu}_{\rho,\lambda}}\ a.s. \tag{12.2}$$

where $w(r,t)$ is the characteristic starting at r and $t(r)$ is the time when $w(r,t)$ and $w(-r,t)$ meet, as defined in Section 2:

$$w(r,t) = \begin{cases} r + (1-2\lambda)t & \text{if } r > 0 \\ r + (1-2\rho)t & \text{if } r < 0 \end{cases}$$

and $t(r) = |r|/(\lambda - \rho)$.

Proof of (12.1). We couple the process $(\eta_t, W(t))$ (the process with only one second class particle located a $W(t)$) and initial measure ν_α with the process $(\sigma_t, \xi_t, X(t))$ with initial measure $\nu_{\rho,\lambda}$, $\rho < \lambda$ and a ξ particle at the origin. Take first $\rho = \alpha$ (hence $\lambda > \alpha$). Initially the two processes are coupled in such a way that $\eta_0 = \sigma_0$ and $W(0) = X(0) = 0$. Since the processes are using the same arrows and $\sigma_t \vdash \xi_t$, we have that $\eta_t \equiv \sigma_t$ for all t. On the other hand,

$$W(t) \geq X(t) \text{ for all } t \text{ almost surely.} \tag{12.3}$$

Equation (12.3) holds at time 0. To prove it for all times it then suffices to show that if $X(t-) = W(t-) = x$, no arrow involving x at time t has the effect that $W(t) < X(t)$. There are the following possibilities:

- - the arrow goes from x to $x+1$ and $\eta_{t-}(x+1) = \sigma_{t-}(x+1) = 1$, then $X(t) = W(t) = x$.
- - the arrow goes from x to $x+1$ and $\eta_{t-}(x+1) = \sigma_{t-}(x+1) = 0$, then $W(t) = x+1$ and $X(t) = x + (1 - \xi_{t-}(x+1))$.
- - the arrow goes from $x-1$ to x and $\eta_{t-}(x-1) = \sigma_{t-}(x-1) = 1$, then $W(t) = x-1$ and $X(t) = x-1$.
- - the arrow goes from $x-1$ to x and $\eta_{t-}(x-1) = \sigma_{t-}(x-1) = 0$, then $W(t) = x$ and $X(t) = x$.

In all cases, after the arrow $W(t) \geq X(t)$. This implies that

$$\liminf_{t\to\infty} \frac{W(t)}{t} \geq \lim_{t\to\infty} \frac{X(t)}{t} = 1 - \rho - \lambda \;\; a.s. \tag{12.4}$$

for all $\lambda > \alpha$, where the identity is the law of large numbers for the shock (11.1).

Now, taking $\lambda = \alpha$ $(\rho < \alpha)$, $\eta_t = \sigma_t + \xi_t$, a similar argument shows that $W(t) \leq X(t)$ almost surely and that

$$\limsup_{t\to\infty} \frac{W(t)}{t} \leq \lim_{t\to\infty} \frac{X(t)}{t} = 1 - \rho - \lambda \;\; a.s. \tag{12.5}$$

Since (12.4) holds for $\rho = \alpha$ and all $\lambda > \alpha$ and (12.5) holds for $\lambda = \alpha$ and all $\rho < \alpha$ this completes the proof of (12.1). ♣

Proof of (12.2). For each r and ε couple $(\eta_t, W(\varepsilon^{-1}r, t))$ with initial measure $\nu_{\rho,\lambda}$ with $(\eta_t, X(t))$, with initial measure $\hat{\nu}_\alpha$. Observe that after $T := \inf\{t : X(t) = W(\varepsilon^{-1}r, t)\}$, we have that $X(t) \equiv W(t)$. Hence one only needs to prove that $\lim_{\varepsilon\to 0}\varepsilon T = t(r)$. This follows from the law of large numbers for $X(t)$ (11.1) and the following law of large numbers for $W(t)$:

$$\lim_{\varepsilon\to 0}\varepsilon W(\varepsilon^{-1}r,\varepsilon^{-1}t) = r + (1-2\alpha)t \quad P_{\nu_\alpha}\ a.s. \tag{12.6}$$

that can be shown as (12.1). ♣

13. Shock Fluctuations

We first show that a perturbation on the initial condition translates as time goes to infinity into a shift of the shock position. For any given configuration η and site $y \in \mathbb{Z}$, let $\eta^{y|i}$ be defined by ($i \in \{0,1\}$):

$$\eta^{y|i}(x) = \begin{cases} \eta(x) & \text{for } x \neq y \\ i & \text{for } x = y. \end{cases}$$

Let $X(\eta, t)$ be the random position of the shock when the initial configuration is η, *i.e.* the position of a second class particle initially at 0. Define $r^+ := (\lambda - \rho)$, $r^- := -(\lambda - \rho)$.

Theorem 13.1. For all $\varepsilon > 0$ it holds

$$\lim_{t\to\infty}\sup_{(r^-+\varepsilon)t<y<(r^+-\varepsilon)t}\left|E_{\nu_{\rho,\lambda}}(X(\eta^{y|0},t) - X(\eta^{y|1},t)) - (\lambda-\rho)^{-1}\right| = 0 \tag{13.2}$$

$$\lim_{t\to\infty}\frac{1}{t}\sum_{y=0}^{r^+t} E_{\nu_{\rho,\lambda}}(X(\eta^{y|0},t) - X(\eta^{y|1},t)) = 1 \tag{13.3}$$

$$\lim_{t\to\infty}\frac{1}{t}\sum_{y=r^-t}^{0} E_{\nu_{\rho,\lambda}}(X(\eta^{y|0},t) - X(\eta^{y|1},t)) = 1 \tag{13.4}$$

Proof. To catch the idea we first prove it for $\rho = 0$. It is convenient to consider an initial configuration ξ distributed according to ν_λ and η_t defined by $\eta_t(x) = \xi_t(x)1\{x \geq X(t)\}$, where $X(t) = X(\eta,t)$. Call $X^i(t) := X(\eta^{y|i},t)$. We have already seen that the two processes will differ at only one site. Let $W(t)$ be the site where the processes with initial configuration $\eta^{y|1}$ and $\eta^{y|0}$ differ by time t. Call $x_n(t)$ the positions at time t of the particles of the process starting at $\xi^{y|0}$, with $x_0(0) = 0$. The shock in both configurations is $X^1(t) = X^0(t) = x_0(t)$ for all times $t \le T_1 := \inf\{t : W(t) < x_0(t)\}$. After T_1, due to the exclusion interaction $X^1(t) = W(t) < x_0(t)$. Call $T_2 := \inf\{t : W(t) = x_{-1}(t)\}$. Now T_2 is finite with probability one. After T_2, $W(t) \equiv x_{-1}(t)$. Since $\nu'_\lambda(x_0 - x_{-1}) = 1/\lambda$ we get the theorem in this case.

When $\rho > 0$ the shock is given also by $x_0(t)$, the 0-th ξ-particle. With the same argument we show that, after T_2, $W(t) \equiv x_{-1}(t)$. Equations (13.3) and (13.4) follow because the expectation of each term is bounded by $\lambda - \rho$. ♣

Define (if the limit exists)

$$D = \lim_{t\to\infty} t^{-1} E_{\nu'_{\rho,\lambda}} (X(t) - vt)^2$$

and

$$\bar{D} := \frac{\rho(1-\rho) + \lambda(1-\lambda)}{\lambda - \rho}.$$

The next result gives $\bar{D}$ as a lowerbound for D.

Theorem 13.5. $D \ge \bar{D}$

The proof of this theorem is a corollary of the next result. In order to state it define

$$D(t) := E_{\nu'_{\rho,\lambda}} (X(t) - vt)^2,$$

$$I(t) := \int d\nu'_{\rho,\lambda}(\eta) E \left(X(\eta, t) - \frac{n_0(\eta, r^+t) - -n_1(\eta, r^-t)}{\lambda - \rho} \right)^2,$$

where $X(\eta, t)$ and $r^\pm$ are defined at the begining of Section 13 and $n_0(\eta, x) := \sum_{y=0}^{x} (1 - \eta(y))$ is the number of empty sites of η between 0 and x and $n_1(\eta, x) := \sum_{y=x}^{0} \eta(y)$ is the number of η particles between the origin and $x < 0$.

Theorem 13.6. The following holds

$$\lim_{t\to\infty} \frac{D(t)}{t} = \bar{D} + \lim_{t\to\infty} \frac{I(t)}{t} \tag{13.7}$$

if the limits exist. If not (13.7) holds with lim substituted by either lim sup or lim inf.

Proof. Summing and substracting vt, $I(t)$ equals

$$\begin{aligned} &\int d\nu'_{\rho,\lambda}(\eta) E \left(X(\eta,t) - vt\right)^2 + \int d\nu'_{\rho,\lambda}(\eta) \left(\frac{n_0(\eta, r^+t)}{\lambda - \rho} - (1-\lambda)t \right)^2 \\ &+ \int d\nu'_{\rho,\lambda}(\eta) \left(\frac{n_1(\eta, r^-t)}{\lambda - \rho} - \rho t \right)^2 \\ &- 2 \int d\nu'_{\rho,\lambda}(\eta) E \left(X(\eta, t) \left(\frac{n_0(\eta, r^+t)}{\lambda - \rho} - (1-\lambda)t - \frac{n_1(\eta, r^-t)}{\lambda - \rho} + \rho t \right) \right) \end{aligned} \tag{13.8}$$

where we have used that $n_0(\eta, r^+t)$ and $n_1(\eta, r^-t)$ are independent under $\nu'_{\rho,\lambda}$. Dividing by t and taking $t \to \infty$, the first term gives $\lim(D(t)/t)$, and the second and third terms give $\bar{D}$. Then it suffices to show that dividing by t and taking $t \to \infty$ the last term equals $-2\bar{D}$. Using the definition of $n_i(.,.)$, the expectation in the last term in (13.8) equals

$$\frac{1}{\lambda-\rho} E\left(\sum_{x=0}^{r^+t} X(\eta,t)(1-\eta(x)-(1-\lambda)) - \sum_{x=r^-t}^{0} X(\eta,t)(\eta(x)-\rho)\right) \tag{13.9}$$

Integrating the first term of (13.9),

$$\begin{aligned}
&-\frac{1}{\lambda-\rho}\int d\nu'_{\rho,\lambda}(\eta)\sum_{x=0}^{r^+t} E\left(X(\eta,t)(\eta(x)-\lambda)\right)\\
&= -\frac{1}{\lambda-\rho}\int d\nu'_{\rho,\lambda}(\eta)\sum_{x=0}^{r^+t}\left[E\left(X(\eta,t)|\eta(x)=1\right)\lambda\right.\\
&\quad \left. -\lambda\big(E\left(X(\eta,t)|\eta(x)=1\right)\lambda + E\left(X(\eta,t)|\eta(x)=0\right)(1-\lambda)\big)\right]
\end{aligned}$$

this equals

$$\begin{aligned}
&= -\frac{1}{\lambda-\rho}\lambda(1-\lambda)\int d\nu'_{\rho,\lambda}(\eta)\\
&\quad\times\sum_{x=0}^{r^+t}\left[E\left(X(\eta,t)|\eta(x)=1\right) - E\left(X(\eta,t)|\eta(x)=0\right)\right] \qquad (13.10)\\
&= -\frac{1}{\lambda-\rho}\lambda(1-\lambda)\int d\nu'_{\rho,\lambda}(\eta)\sum_{x=0}^{r^+t} E\left(X(\eta^{x|1},t) - X(\eta^{x|0},t)\right)
\end{aligned}$$

Dividing by t and taking the limit as $t \to \infty$ of the first term of (13.9), we get using (13.3) on (13.10) that

$$-\frac{1}{\lambda-\rho}\lim_{t\to\infty}\frac{1}{t}\int d\nu'_{\rho,\lambda}(\eta)\left(\sum_{x=0}^{r^+t} X(\eta,t)(\eta(x)-\lambda)\right) = \frac{\lambda(1-\lambda)}{\lambda-\rho}$$

and analogously using (5.4),

$$-\frac{1}{\lambda-\rho}\lim_{t\to\infty}\frac{1}{t}\int d\nu'_{\rho,\lambda}(\eta)\sum_{x=r^-t}^{0} X(\eta,t)(\eta(x)-\rho) = \frac{\rho(1-\rho)}{\lambda-\rho}$$

This implies the Theorem. ♣

Remarks. From Theorem 13.6 we conclude that the diffusion coefficient of the shock is the same as the conjectured diffusion coefficient if and only if the position of the shock at time t is given –in the scale $\sqrt{t}$– by $(\lambda - \rho)^{-1}$ times the number of holes between 0 and r^+t minus the number of particles between 0 and r^-t. In any case, $I(t)$ is non negative, then $\bar{D}$ is always a lower bound and this proves Theorem 13.5. When $\rho = 0$, $X(t)$ has the distribution of a plain tagged particle in the simple exclusion process with density λ In this case it is known that $D := \lim_{t\to\infty} t^{-1}E(X(t) - EX(t))^2 = \bar{D} = (1-\lambda)$ (Corollary 7.2). This implies that $\lim_{t\to\infty} I(t)/t = 0$, hence in the scale $\sqrt{t}$ the position of $R(t)$ is determined by the initial configuration in the sense discussed above. This was proved by Gärtner and Presutti [gp] using a different method.

We finish this section by mentioning a couple of open problems.

Prove that there exists a microscopic shock in more dimensions or for a jump function that allows to go further than to the nearest neighbors. The argument of this approach does not work even for the case of two parallel lines with a symmetric dynamics for jumps between the two lines and asymmetric jumps inside each line. Landim [la1] proved the existence of the hydrodynamical limit in high dimensions for some initial conditions.

General initial conditions. Benassi, Fouque, Saada and Vares [bfsv] have proved that the hydrodynamical limit can be taken when the initial profile is monotone and the jump probabilities are more general. Prove the hydrodynamical limit for any initial profile.

Fluctuations. It is expected that the fluctuations arround the deterministic hydrodynamic limit depend on the initial configuration as the fluctuations of the shock do. In the case $\rho = 0$ this has been studied by [bf2].

14. The Nearest Neighbor Asymmetric Simple Exclusion Process. General Case.

Almost all the results described above have been proven for the process where the particles can also jump backwards. In this case one assumes that the particles jump at rate p to the right nearest neighbor and with rate q to the left one. We assume $p + q = 1$ and $p > q$.

The graphical representation works in the same way. The only difference is that now we have different Ppp per bond $(x, x+1)$. One associated to arrows going from x to $x+1$ with rate p and the other associated to arrows going from $x+1$ to x at rate q. As before we can couple two or more versions of the process making the different versions to follow the same arrows.

The set of extremal invariant measures is $\{\nu_\alpha : 0 \le \alpha \le 1\} \cup \{\nu^{(n)} : n \in \mathbb{Z}\}$. The measure $\nu^{(n)}$ concentrates on the set $\mathbf{A}_n := \{\eta : \sum_{x\ge 0}(1-\eta(x)) - \sum_{x<0}\eta(x) = n\}$. They are defined by $\nu^{(n)} := \nu^{[k]}(.|\mathbf{A}_n)$ for all k, where $\nu^{[k]}$ is also product, have marginals

$$\nu^{[k]}(\eta(x)) = \frac{(p/q)^{x-k}}{1-(p/q)^{x-k}} \tag{14.1}$$

and are even reversible for the process. They approach exponentially fast the densities 0 and 1 to the left and right of the origin respectively, so that, under $\nu^{(n)}$, the origin is a shock for $\rho = 0$ and $\lambda = 1$. Now we construct a reversible measure for the process $(\eta_t, R(t))$, where $R(t)$ stands for the position of the second class particle at time t. Put a second class particle at i with probability

$$m(i) := M(p/q)\left((1+(p/q)^{i-\frac{1}{2}})(1+(q/p)^{i+\frac{1}{2}})\right)^{-1} \tag{14.2}$$

where $M(p/q)$ is a normalizing constant making $\sum m(i) = 1$. For the other sites decide that a η particle is present at site j with probability

$$m(j|i) = \begin{cases} (p/q)^{j-\frac{1}{2}}/(1+(p/q)^{j-\frac{1}{2}}) & \text{if } j < i \\ (p/q)^{j+\frac{1}{2}}/(1+(p/q)^{j+\frac{1}{2}}) & \text{if } j > i \end{cases}$$

independently of everything. Since the measure concentrates on a denumerable state space, that statement follows from a routine computation. It is clear from (14.2) that the second class particle will remain tight, hence it also is a shock in this case.

In the case $0 \le \rho < \lambda \le 1$ one follows the same steps as in the case $p = 1$. So we construct a process $(\sigma_t, \xi_t, X(t))$, with $\sigma_t \vdash \xi_t$ and $X(t)$ is a tagged ξ particle. Then one calls $x_i(t)$ the positions of the ξ particles, assuming that $x_0(t) \equiv X(t)$. Finally one chooses the i-th ξ_t particle to be $R(t)$ with probability $m(i)$ and label the others ξ particles independently in the following way: the j-th particle is labeled γ with probability $m(j|i)$ otherwise it is labeled ζ. The remarkable property of this distribution is that the labeling remains invariant for later times. Hence the density of γ particles vanishes to the left of $X(t)$ and the density of ζ particles vanishes to the right of $X(t)$ exponentially fast. Furthermore $X(t) - R(t)$ remains tight. Since our original process can be recovered by writing $\eta_t = \sigma_t + \gamma_t$ as before one can prove that either $X(t)$ or $R(t)$ are microscopic shocks. The advantage of $R(t)$ is that it can be defined directly as a second class particle with respect to η_t, while for defining $X(t)$ one needs to use the process (σ_t, ξ_t).

With this observation and a little care one can prove all the other results. We refer to [fks] and [f2] for details.

15. The Weakly Asymmetric Simple Exclusion Process

We have studied the hydrodynamical limit of a process by rescaling time and the space in a convenient way. The rescaling is done with a parameter ε. Another possibility is to consider not a single process, but a family of processes depending on the parameter ε. Indeed this is what is done to derive the (viscous) Burgers equation

$$\frac{\partial u}{\partial t} - \frac{\partial^2 u}{\partial r^2} + \frac{\partial}{\partial r}(u(1-u)) = 0 \tag{15.1}$$

This kind of limit is called kinetic. In this case the family consists of asymmetric simple exclusion processes as defined in the previous section with $p = \frac{1+\varepsilon}{2}$ and $q = \frac{1-\varepsilon}{2}$. Since the asymmetry $p - q$ goes to 0 with the scaling parameter ε, the resulting (family of) process(es) is called weakly asymmetric. It was introduced by De Masi, Presutti and Scacciatelli [dps] and studied by Gartner [g], Dittrich [d] Gartner and Dittrich [gd] and Ferrari, Kipnis and Saada [fks]. The results are quite complete and reinforce the conjectures for the simple exclusion process we did in the previous sections.

Let L^ε be defined by $L^\varepsilon = (1-\varepsilon/2)L_0 + \varepsilon L$, where L is the generator of the simple exclusion process defined in (3.2) and

$$L_0 f = \sum_x \tfrac{1}{2}[f(\eta^{x,x+1}) - f(\eta)]$$

is the generator of a symmetric exclusion process. Denoting by $S^\varepsilon(t)$ the semigroup corresponding to the generator L^ε, and by ν^ε a family of product mesures with marginals given by $\nu^\varepsilon(\eta(\varepsilon^{-1}r)) = u_0(r)$ the kinetic limit is given by

$$\lim_{\varepsilon\to 0} \nu^\varepsilon S^\varepsilon(\varepsilon^{-1}t)\tau_{[\varepsilon^{-1/2}r]} f = \nu_{u(r,t)} f$$

(local equilibrium), where $u(r,t)$ is the solution of (15.1) with inital condition $u(r,0) = u_0(r)$. This result was proven by De Masi, Presutti and Scacciatelli [dps] and by Gartner [g]. Notice that the scaling is different from the one we used to derive the unviscous Burgers equation. To obtain the Laplacian $\partial^2 u/\partial^2 r$ one needs to scale space as square root of time. Since we are looking at time ε^{-1} and L is multiplied by ε, the asymmetry is not rescaled but it appears anyway in the macroscopic limit as a transport term. The same authors proved the law of large numbers for the density fields: Let Φ be a smooth function with compact support. Then, calling η_t^ε the process with generator L^ε,

$$\lim_{\varepsilon\to 0} \varepsilon \sum_{x\in\mathbb{Z}} \Phi(\varepsilon x)\eta^\varepsilon_{\varepsilon^{-1}t}(x) = \int_{\mathbb{R}} \Phi(r)u(r,t)dr$$

$P^{\varepsilon}_{\nu_\varepsilon}$ almost surely. The stationary case was study by Ferrari, Kipnis and Saada [fks]: For each ε there exists a position $X(t)$ such that the process as seen from $X(t)$ has a law μ^{ε} with the property that for all ε, $\mu^{\varepsilon} \sim \nu_{\rho,\lambda}$. Furthermore

$$\lim_{\varepsilon\to 0} \mu^{\varepsilon}(\eta(\varepsilon^{-1/2}r)) = u(r)$$

where

$$u(r) := \rho + \frac{\lambda-\rho}{1+e^{-2r(\lambda-\rho)}}$$

is the stationary travelling wave solution of the Burgers equation (15.1) with asymptotic densities ρ and λ. Also the density fields converge and the hydrodynamical limit is performed for this family of initial measures. The equilibrium case in a finite macroscopic box was studied by De Masi, Ferrari and Vares [dfv].

A stronger result has been proved by Dittrich [d], who exibits a function ξ_t of the initial configuration η_0, such that for any test function ψ,

$$\lim_{\varepsilon\to 0} \varepsilon \sum_x (\eta^{\varepsilon}_{\varepsilon^{-1}t}(x) - \xi_{\varepsilon^{-1}t}(x))\psi(\varepsilon x) = 0$$

In other words, in that scale the motion is determined by the initial configuration. The approach also allows to study the shock wave case. In this case he proves that the shock fluctuates as a Brownian motion with diffusion coefficient given by $\bar{D}$, defined in Section 13 and that these fluctuations depend only on the initial configuration.

An interesting problem is to decide what happens with the second class particle in this limit. Calling $R(t)$ the position of the second class particle at time t, with $R(0) = 0$, the rescaling of (14.2) gives

$$\lim_{\varepsilon\to 0} P_{\mu^{\varepsilon}}(R^{\varepsilon}(t) - vt \leq \varepsilon^{-(1/2)}r) = M\int_{-\infty}^{r} \frac{\lambda-\rho}{(1+\exp(-2s(\lambda-\rho)))(1+\exp(2s(\lambda-\rho)))}ds$$

where $M = \lim_{\varepsilon\to 0} M((\frac{1}{2}+\varepsilon)/(\frac{1}{2}-\varepsilon))$ is the limit of the normalizing constants of (14.2). We conjecture that the limiting motion of the second class particle is a Ornstein Uhlenbeck process with an appropriate drift with stationary measure given above.

De Masi, Presutti and Scacciatelli [dps] have studied the fluctuation fields and proved that they converge to a generalized Ornstein Uhlenbeck process. Ravishankar [ra] studied the d-dimensional case.

16. The Boghosian Levermore Cellular Automaton

In this section we review recent results of Ferrari and Ravishankar [fr] on a deterministic version of a probabilistic cellular automaton introduced by Boghosian and Levermore [bl]. This is a dynamical system with random initial condition. The simplicity of the model allows to prove all the known results and conjectures for the asymmetric simple exclusion process.

A configuration of the one dimensional Boghosian Levermore Cellular automaton (BCLA) is an arrangement of particles with velocities +1 and −1 on $\mathbb{Z}$, satisfying the exclusion condition that there is at most one particle with a given velocity (+1 or −1) at each site. We denote a configuration by $\eta \in \{0,1\}^{\mathbb{Z}\times\{-1,+1\}} := \mathbf{X}$, the state space. If $\eta(x,s) = 1$ we say that there is a particle with velocity s at site x, where $x \in \mathbb{Z}$ and $s \in \{-1,+1\}$.

Dynamics. The time is discrete and the dynamics is given in two steps:

1. Collision: for a given η let $C\eta$ be the configuration

$$C\eta(x,+1) = 1\{\eta(x,1) + \eta(x,-1) \geq 1\} = \max\{\eta(x,1), \eta(x,-1)\}$$
$$C\eta(x,-1) = 1\{\eta(x,1) + \eta(x,-1) = 2\} = \min\{\eta(x,1), \eta(x,-1)\}$$

In other words, if there is no particle or two particles at x, then nothing happens. If there is only one particle at x, then this particle adopts velocity 1.

2. Advection. This part of the dynamics moves each particle along its velocity to a neighboring site in unit time. The operator A is defined by

$$A\eta(x,s) = \eta(x-s,s).$$

Defining $T := AC$, the dynamics is given by

$$\eta_{t+1} = T\eta_t$$

We say that a measure μ on $\mathbf{X}$ is stationary for the process if $\mu T = \mu$. Cheng, Lebowitz and Speer [cls] have noticed that this dynamics acts independently in the space-time sublattices $\{(x,t) : x+t \text{ is even}\}$ and $\{(x,t) : x+t \text{ is odd}\}$. Any translation invariant measure concentrating in one of the sets $\{\eta : \eta(x,1) = 1\}$, $\{\eta : \eta(x,-1) = 0\}$ is stationary for the process. Also there are non translation invariant measures, the measures ν^n giving mass 1/2 to η^n and $T\eta^n$ where $\eta^n(x,\pm 1) = 1\{x \geq n\}$. Observe that the configuration η^n is two steps invariant (i.e. $\eta^n = T^2\eta^n$). The problem of determining if the measures introduced above are sufficient to describe the set of all invariant measures for the process is open. We remark that in contrast with simple exclusion, there are stationary translation invariant ergodic measures that are not product measures.

The hydrodynamical limit has been performed for initial product measures with densities 0 and ρ for particles with velocity -1 and $+1$ respectively at negative sites, and with densities λ and 1 for particles with velocities -1 and $+1$, respectively to the right of the origin. Hence the particle density per site to the left of the origin is ρ and to its right is $1+\lambda$. The limiting density per site satisfies the equation

$$\frac{\partial u}{\partial t} + F(u) = 0$$

where

$$F(u) = u1\{0 \leq u \leq 1\} + (2-u)1\{1 \leq u \leq 2\}$$

This equation has only two characteristics, 1 and -1 according to the density being bigger or smaller than 1. The definition of the microscopic shock is the same as for the simple exclusion process. Just take two configurations that differ at only one velocity at only one site. At latter times they will also differ at only one site that we call second class particle, as it gives priority to other particles that attempt to occupy its place.

When the initial configuration is taken from the product measure described above, it turns out that the position of the second class particle can be expressed as a sum of a random number of independent and identically distributed random variables, each of which is a difference of two geometric random variables. Furthermore the number of summands is independent of the summands. Hence, as a corollary of this result we can prove laws of large numbers and central limit theorems for the position of the shock. Another result is that the position of the shock at any given time is independent of the configuration at that time as seen from the shock. This gives a way to prove the hydrodynamical limit described above.

The results can be extended to initial measures with more than one step and to decreasing profiles and also to the probabilistic cellular automata when the C rule is applied with probability p and is not applied with probability $1-p$ [frv]. The weakly asymmetric case was studied by [lop].

17. Other Cellular Automata

In this section we show that the BLCA is isomorphic to two simple exclusion type of automata in $\{0,1\}^{\mathbb{Z}}$ and a type of sand-pile automaton.

The asymmetric simple exclusion cellular automaton. For a given configuration $\xi \in \{0,1\}^{\mathbb{Z}}$, define $B_1\xi$ as the configuration

$$\begin{aligned} B_1\xi(2z+1) &= 1\{\xi(2z)+\xi(2z+1) \geq 1\} = \max\{\xi(2z), \xi(2z+1)\} \\ B_1\xi(2z) &= \xi(2z)+\xi(2z+1) - 1\{\xi(2z)+\xi(2z+1) \geq 1\} \\ &= \min\{\xi(2z), \xi(2z+1)\} \end{aligned}$$

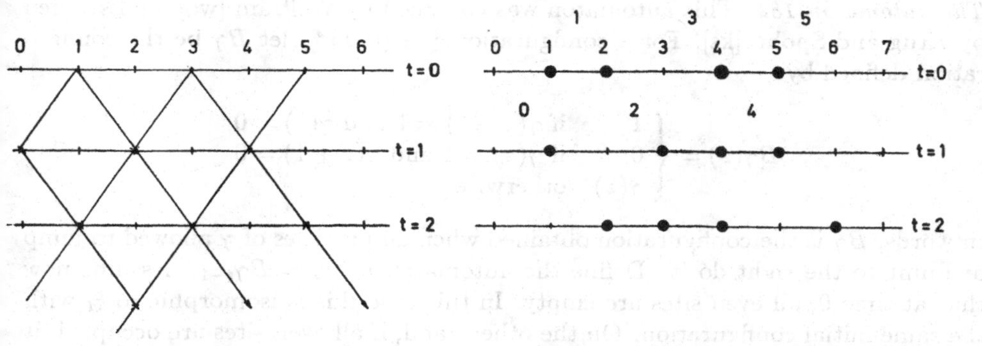

Figure 9. Isomorphism between the BLCA and the ASECA.

and $B_2\xi$ as the configuration

$$B_2\xi(2z) = 1\{\xi(2z-1) + \xi(2z) \geq 1\}$$
$$B_2\xi(2z-1) = \xi(2z-1) + \xi(2z) - 1\{\xi(2z-1) + \xi(2z) \geq 1\}$$

Now define the asymmetric simple exclusion cellular automaton (ASECA):

$$\xi_{2t} = B_1\xi_{2t-1} \text{ and } \xi_{2t+1} = B_2\xi_{2t} \tag{17.1}$$

In words, at even times, all particles occupying even sites that can jump to the right (i.e. that the succesive odd site is empty), jump to the right. At odd times, the particles occupying the odd sites do the same.

We prove that this is isomorphic to a subsystem of the BLCA. As observed before, the BLCA consists in two independent subsystems: $\{\eta(x,s,t) : x+t \text{ odd }\}$ and $\{\eta(x,s,t) : x+t \text{ even }\}$. Consider the subsystem $\{\eta(x,s,t) : x+t \text{ odd }\}$ and define the configuration ξ_t by

$$\xi_t(2x+1) = \begin{cases} \eta(2x+1,-1,t) & \text{for } t \text{ even} \\ \eta(2x,1,t) & \text{for } t \text{ odd} \end{cases} \tag{17.2.a}$$

and

$$\xi_t(2x) = \begin{cases} \eta(2x-1,1,t) & \text{for } t \text{ even} \\ \eta(2x,-1,t) & \text{for } t \text{ odd} \end{cases} \tag{17.2.b}$$

We have the following result which proof is straightforward (see Figure 9)

Lemma 17.3. The transformation (17.2) defines an isomorfism between the subsystem $\{\eta(x,s,t) : x+t \text{ odd }\}$ and $\xi_t, t \in \mathbb{Z}$, such that ξ_t is the asymmetric simple exclusion cellular automaton, with distribution described by (17.1).

The automaton 184. This automaton was classified by Wolfram [wo], and studied by Krug and Spohn [ks]. For a configuration $\gamma \in \{0,1\}^{\mathbb{Z}}$, let $B\gamma$ be the configuration defined by

$$B\gamma(z) = \begin{cases} 1 & \text{if } \gamma(z-1) = 1 \text{ and } \gamma(z) = 0 \\ 0 & \text{if } \gamma(z) = 1 \text{ and } \gamma(z+1) = 0 \\ \gamma(z) & \text{otherwise} \end{cases}$$

In words, $B\gamma$ is the configuration obtained when all particles of γ allowed to jump one unit to the right do it. Define the automaton by $\gamma_t = B\gamma_{t-1}$. Assume now that at time 0, all even sites are empty. In this case this is isomorphic to ξ_t with the same initial configuration. On the other hand, if all even sites are occupied, it is also isomorphic to ξ_t. Nevertheless, for other configurations this system is not isomorphic, presenting a richer structure. But for those tipe of initial conditions the results proved for the BLCA hold.

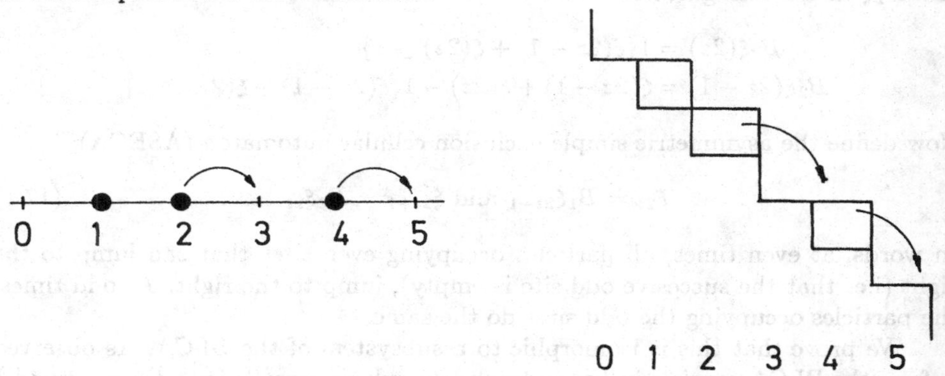

Figure 10. Isomorphism between the automaton 184 and a sand-pile.

Sand piles. We consider now an infinite version of an automaton introduced by Bak [b] and studied by Goles [g]. Consider a process ζ_t on the state space $\{\zeta \in \mathbb{Z}^{\mathbb{Z}} : \zeta(x) \geq \zeta(x+1)\}$. Define $C : \zeta \mapsto C\zeta$ as follows

$$C\zeta(x) = \zeta(x) + 1\{\zeta(x-1) - \zeta(x) \geq 2\} - 1\{\zeta(x) - \zeta(x+1) \geq 2\}$$

In words, we can think that at each integer there is a pile of grains of sand. At each time each pile is ready to give one of its grains to its right neighboring pile. But this only happens if after that the receiving pile is not higher than the one that is giving the grain. This operations are all done in parallel. The automaton is defined by

$$\zeta_t = C\zeta_{t-1}$$

It turns out that for some initial configurations, this automaton is isomorphic to the Automaton 184 described above. This has been established in [fgv]. Let γ be a configuration of $\{0,1\}^{\mathbb{Z}}$ and define $\zeta = \zeta(\gamma)$ as the configuration

$$\zeta(x) = -x + \gamma(x) \tag{17.4}$$

Then it is easy to see that

$$C\zeta(x) = -x + B\gamma(x)$$

so that we get that $\zeta_t(x) = -x + \gamma_t(x)$ for all t, all x. This implies that all the results for the hydrodynamics and shocks hold for this model for this kind of initial conditions. Other initial conditions are under investigation by [fgv].

It is not hard to see that a particle system can be constructed using the same law. Assume that at each site we have a Poisson point process of rate one, independent of everything. When the clock rings at site x, it attempts to give a grain to site $x+1$, but it does so only if $\zeta(x) - \zeta(x+1) \geq 2$. This graphical construction is useful to show the existence of such a particle system because as in the simple exclusion case, at each given time we can make a (random) partition of the space in finite boxes that do not interact among them. One can show that the transformation (17.4) makes this system isomorphic to the simple exclusion process.

Acknowledgments

I thank Antonio Galves, Eric Goles, Claude Kipnis, Joel Lebowitz, Errico Presutti, K. Ravishankar, Ellen Saada, Eugene Speer, Herbert Spohn and Maria Eulalia Vares for enjoyable discussions. This notes started as a series of lectures I gave at Universities of l'Aquila, Roma I and II in 1989. I thank the warm hospitality of my italian colleages. I would like also to thank to Servet Martínez and Eric Goles for their hospitality during the FIESTA 90 School. This work is partially supported by FAPESP, CNPq and agreement CNPq-NSF.

References

[abl] E. D. Andjel, M. Bramson, T. M. Liggett (1988). Shocks in the asymmetric simple exclusion process. *Probab. Theor. Rel. Fields* **78** 231-247.

[ak] E. D. Andjel, C. Kipnis (1984). Derivation of the hydrodynamical equations for the zero-range interaction process. *Ann. Probab.* **12** 325-334.

[av] E. D. Andjel, M. E. Vares (1987). Hydrodynamic equations for attractive particle systems on $\mathbb{Z}$. *J. Stat. Phys.* **47** 265-288.

[b] P. Bak, Ch. Tang, K. Wiesenfeld (1988). Self-Organized criticallity. *Phys. Rev.* **A, 38** (1) 364-373.

[bf1] A. Benassi, J-P. Fouque (1987). Hydrodynamical limit for the asymmetric simple exclusion process. *Ann. Probab.* **15** 546-560.

[bf2] A. Benassi, J-P. Fouque (1990). Fluctuation field for the asymmetric simple exclusion process. Preprint.

[bfsv] A. Benassi, J-P. Fouque, E. Saada, M. E. Vares (1991) Asymmetric attractive particle systems on Z: hydrodynamical limit for monotone initial profiles. *J. Stat. Phys.*

[bg] F. Bertein, A. Galves (1977) Comportement asymptotique de deux marches aleatoires sur Z qui interagissent par exclusion. *C.R. Acad. Sci. Paris A,* **285,** 681-683.

[bl] B. M. Boghosian, C. D. Levermore (1987). A cellular automaton for Burgers' equation. *Complex Systems* **1** 17-30.

[bcfg] C. Boldrighini, C. Cosimi, A. Frigio, M. Grasso-Nunes (1989). Computer simulations of shock waves in completely asymmetric simple exclusion process. *J. Stat. Phys.* **55,** 611-623.

[b] M. Bramson (1988) Front propagation in certain one dimensional exclusion models. *J. Stat. Phys.* **51,** 863-869.

[B] L. Breiman (1968). Probability. Addison-Wesley, Reading, Massachusetts.

[cls] Z. Cheng, J. L. Lebowitz, E. R. Speer (1990). Microscopic shock structure in model particle systems: the Boghosian Levermore revisited. Preprint.

[df] A. De Masi, P. A. Ferrari (1985) Self diffusion in one dimensional lattice gases in the presence of an external field. *J. Stat. Phys.* **38,** 603-613.

[dfv] A. De Masi, P. A. Ferrari, M. E. Vares (1989) A microscopic model of interface related to the Burgers equation. *J. Stat. Phys.* **55,** 3/4 601-609.

[dkps] A. De Masi, C. Kipnis, E. Presutti, E. Saada (1988). Microscopic structure at the shock in the asymmetric simple exclusion. *Stochastics* **27,** 151-165.

[dps] A. De Masi, E. Presutti, E. Scacciatelli (1989), The weakly asymmetric simple exclusion process. *Ann Inst. Henry Poincaré,* **25,** 1:1-38.

[d] P. Dittrich (1989) Travelling waves and long time behaviour of the weakly asymmetric exclusion process. *Probab. Theor. Related Fields.*

[f1] P. A. Ferrari (1986). The simple exclusion process as seen from a tagged particle. *Ann. Probab.* **14** 1277-1290.

[f2] P. A. Ferrari (1990) Shock fluctuations in asymmetric simple exclusion. To appear in *Probab. Theor. Related Fields.*

[fgv] P. A. Ferrari, E. Goles, M. E. Vares (1990). In preparation.

[fks] P. A. Ferrari, C. Kipnis, E. Saada (1991) Microscopic structure of travelling waves for asymmetric simple exclusion process. *Ann. Probab.* **19** 226-244.

[fr] P. A. Ferrari, K. Ravishankar (1990). Shocks in asymmetric exclusion cellular automata. Preprint IME-USP. Submitted *Ann. Appl. Probab.*

[frv] P. A. Ferrari, K. Ravishankar, M. E. Vares (1990). Shocks in asymmetric exclusion probabilistic cellular automata. In preparation.

[fo] J-P. Fouque (1989). Hydrodynamical behavior of asymmetric attractive particle systems. One example: one dimensional nearest neighbors asymmetric simple exclusion process. Preprint.

[g] J. Gärtner (1988). Convergence towards Burger's equation and propagation of chaos for weakly asymmetric exclusion processes. *Stochastic Process Appl.* **27**, 233-260.

[gp] J. Gärtner, E. Presutti (1989). Shock fluctuations in a particle system. CARR Reports in Math. Physics 1/89.

[go] E. Goles (1990). Sand Piles, combinatorial games and cellular automata. Preprint.

[h1] T. E. Harris (1967). Random measures and motions of point processes. *Z. Wahrsch. verw. Gebiete.* **9** 36-58.

[h2] T. E. Harris (1978). Additive set-valued Markov processes and graphical methods. *Ann. Probab.* **6** 355-378.

[k] C. Kipnis (1986). Central limit theorems for infinite series of queues and applications to simple exclusion. *Ann. Probab.* **14** 397-408.

[ks] J. Krug, H. Spohn (1988). Universality classes for deterministic surface growth. *Phys. Review* **A, 38** 4271-4283.

[la1] C. Landim (1989). Hydrodynamical equation for attractive particle systems on Z^d. Preprint.

[la2] C. Landim (1990). Hydrodynamical limit for asymmetric attractive particle systems in $\mathbb{Z}^d$. Preprint.

[lax] P. D. Lax (1972). The formation and decay of shock waves. *Amer. Math. Monthly* (March).

[lop] J. L. Lebowitz, V. Orlandi, E. Presutti (1989). Convergence of stochastic cellular automaton to Burger's equation. Fluctuations and stability. Preprint.

[lps] J. L. Lebowitz, E. Presutti, H. Spohn (1988). Microscopic models of hydrodynamical behavior. *J. Stat. Phys.* **51,** 841-862

[l] T. M. Liggett (1976). Coupling the simple exclusion process. *Ann. Probab.* **4** 339-356.

[L] T. M. Liggett (1985). *Interacting Particle Systems.* Springer, Berlin.

[ps] S. C. Port, C. J. Stone (1973). Infinite particle systems. *Trans. Amer. Math. Soc.* **178** 307-340.

[s1] E. Saada (1987). A limit theorem for the position of a tagged particle in a simple exclusion process. *Ann. Probab.* **15** 375-381.

[s2] E. Saada (1988). Mesures invariantes pour les sistèmes à une infinité de particules linéaires à valeurs dans $[0,\infty[^S$.

[spi] F. Spitzer (1970). Interaction of Markov processes. *Adv. Math.*, **5** 246-290.

[S] H. Spohn (1989). *Large Scale Dynamics of Interacting Particles. Part B: Stochastic Lattice Gases.* Preprint.

[ra] K. Ravishankar (1990) Preprint.

[rb] M. Rosenblatt (1967). Transition probability operators. *Proc. Fifth Berkeley Symp. Math. Stat. Prob.* **2** 473-483.

[r] H. Rost (1982) Nonequilibrium behavior of a many particle process: density profile and local equilibrium. *Z. Wahrsch. verw. Gebiete,* **58** 41-53.

[vb] H. Van Beijeren (1991) Fluctuations in the motions of mass and of patterns in one-dimensional driven diffusive systems. *J. Stat. Phys.*

[wa] J. Walker (1989). How to analyze the shock waves that sweep through expressway traffic. *Scientific American* August 1989:84-87.

[w] D.Wick (1985). A dynamical phase transition in an infinite particle system. *J. Stat. Phys.* **38** 1015-1025.

[wo] S. Wolfram (1983). *Rev. Mod. Phys.* **55** 601.

AUTOMATA NETWORKS STRATEGIES FOR OPTIMIZATION PROBLEMS

E. GOLES
S. MARTINEZ
Universidad de Chile
Facultad de Ciencias Físicas y Matemáticas
Departamento de Ingeniería Matemática
Casilla 170 correo 3
Santiago, Chile

1. Introduction

The goal of this work is to show that a positive automata iteration is the natural hill-climbing strategy of a subnorm or norm optimization problem.

The initial idea of associating automata networks with optimization problems arises in the context of computing fundamental states of spin-glasses by Monte-Carlo methods [MR]. In this method, known as the Metropolis algorithm, a new configuration C' is obtained from the initial one C, by local variation. If the cost function decreases, the configuration C' is accepted and the process continues. If not, C' is accepted with probability $\exp(-\beta\frac{\Delta E}{T})$, where T is a temperature parameter and ΔE the variation of the cost function. Recall that this procedure introduces a local dynamics of an automata network. The zero temperature limit, $T = 0$, corresponds to a classical hill-climbing local deterministic strategy: the configuration C' is accepted if and only if $E(C') < E(C)$. This kind of approach was extensively used for computing fundamental states of spin-glasses for tridimensional signed lattices where the optimization problem is NP-complete [B,GJ].

More recently Kirkpatrick et al. [K] have introduced the simulated annealing, which is an stochastic relaxation method based on the Metropolis algorithm. It was proven (see [LA]) that if the temperature parameter decreases slow enough, then the relaxation converges to the optimum value. The simulated annealing presents the problem of being extremely time consuming. Furthermore, it is only needed, in several optimization problems, a "good" solution which does not necessarily means the optimum one. In this context, deterministic search derived from mean-field approximation of the stochastic relaxation, which are crude versions of previous algorithms, can be used [PA].

E. Goles and S. Martínez (eds.), Statistical Physics, Automata Networks and Dynamical Systems, 65–87.

Since the paper of Hopfield and Tank [HT], several authors have used the neural network approach as hill-climbing algorithms for NP-complete optimization problems [PA]. The main idea is based upon the fact that the sequential and the parallel updating of symmetric neural networks is driven by Ising-like Hamiltonians. More precisely, consider the sequential updating of a neural network,

$$x_i' = \text{sign}(\sum_{j=1}^{n} a_{ij}x_j - b_i) \quad i = 1, ..., n, \quad x_i \in \{-1, 0, +1\}. \tag{1.1}$$

If the interactions are symmetric: $a_{ij} = a_{ji}$ for any i, j, and the diagonal is null $a_{ii} = 0$ for any i, then the Hamiltonian

$$H(x) = -\frac{1}{2}\sum_{i=1}^{n}\sum_{j=1}^{n} a_{ij}x_ix_j + \sum_{i=1}^{n} b_ix_i \tag{1.2}$$

decreases with the dynamics. In fact if we are updating the i-th unit then $x_k' = x_k$ for $k \neq i$ and

$$H(x') - H(x) = -(x_i' - x_i)(\sum_{j=1}^{n} a_{ij}x_j - b_i) \leq 0.$$

On the other hand, for the parallel updating

$$x_i(t+1) = \text{sign}\,(\sum_{j=1}^{n} a_{ij}x_j(t) - b_i) \tag{1.3}$$

of a symmetric neural network, the energy function

$$E(x(t)) = -\sum_{i=1}^{n}\sum_{j=1}^{n} a_{ij}x_i(t)x_j(t-1) + \sum_{i=1}^{n} b_i(x_i(t) + x_i(t-1)) \tag{1.4}$$

is decreasing.

The form of these functionals H and E allow us to show that [H,GM]:

- the sequential updating converges to fixed points which are local minima of H and,

- the parallel updating converges to 2-cycles, which are local minima of E.

Then, symmetric neural networks evolve to local minima of combinatorial optimization problems with quadratic symmetric cost functions. Hence, neural networks dynamics constitutes a hill-climbing strategy.

Furthermore, global minima are stationary points of the dynamics, but in order to reach them we must choose the initial values in the basin of attraction of the global minima. Recently, a new method [HT,PA] consisting in the replacement of the sign function by a smooth one, for instance by a parametrized sigmoid, has permited to get better solutions than those obtained by using discrete transition functions.

Let us detail the strategy for associating neural networks to specific problems. We shall make it for the graph bisection problem which appears in the practical context of microcircuit'design and it is a classical NP-complete problem.

Consider an undirected and boucle free graph $G = (V, E)$, where V is the set of vertices and E is the set of non-oriented edges. We assume that the number of vertices $n = |V|$ is even. The graph bisection problem consists in splitting the set V into two sets V_1, V_2 of equal number of vertices in such a manner that the number of edges between V_1 and V_2 is minimized. Denote $\alpha(V_1, V_2) = |\{(i,j) \in E : i \in V_1, j \in V_2\}|$, we call this quantity the cutsize. Then the combinatorial optimization problem consists in:

$$(P1) \quad \begin{array}{l} \min \alpha(V_1, V_2) \\ \text{s.t. } \{V_1, V_2\} \text{ is a partition of } V \text{ verifying } |V_1| = |V_2| \end{array}$$

In order to approach this problem by a neural network we associate each vertex $i \in V = V_1 \cup V_2$ with a binary variable $x_i \in \{-1, 1\}$. We set $x_i = 1$ if $i \in V_1$ and $x_i = -1$ if $i \in V_2$. We also take $a_{ij} = 1$ if $(i,j) \in E$, 0 otherwise. Since the graph is undirected and boucle free, the matrix is symmetric with null diagonal. By definition:

$$a_{ij}x_ix_j = \begin{cases} +1 & \text{if } i,j \text{ belong to the same } V_i \text{ and } (i,j) \in E \\ 0 & \text{if } (i,j) \notin E \\ -1 & \text{if } i,j \text{ belong to different } V_i \text{ and } (i,j) \in E \end{cases}$$

On the other hand, the restriction $|V_1| = |V_2|$ is equivalent to $\sum_{j=1}^{n} x_j = 0$.

In this context we associate to the bisection graph problem the near cost function:

$$(P2) \qquad H(x) = -\frac{1}{2}\sum_{i=1}^{n}\sum_{j=1}^{n} a_{ij}x_ix_j + \frac{\lambda}{2}(\sum_{i=1}^{n} x_i)^2$$

where $\lambda > 2$ is a parameter. So, the minimization of $H(x)$ is a trade off between the restriction feasibility $\sum_{i=1}^{n} x_i = 0$ and the minimality of the cutsize. We may write $H(x)$ as: $H(x) = -\frac{1}{2}\sum_{i=1}^{n}\sum_{j=1}^{n} b_{ij}x_ix_j$, where the connections are $b_{ij} = a_{ij} - \lambda$.

So, with this problem we associate the updating rule $x_i' = \text{sign}\,(\sum_{j=1}^{n} b_{ij}x_j)$, which converges to local minima of (P2), so of (P1). On the other hand, the optimal set of the partition graph problem are minima of (P2). However, the hill-climbing strategy iterated in a sequential or parallel may converge to very high local minima. In several works [HT,PA] this problem is solved by replacing the local function sign by a sigmoidal function parametrized in such a way that it converges to the sign function. It has been found that the stationary points of the sigmoidal iteration, converges with the parameter, to better minima. Several other combinatorial optimization problems have been treated in this way. But some quantities must be often introduced, like λ in previous example, which cannot be computed exactly.

So, the network approach remains as an approximation method which is a good one, only if there exists an appropriated calculation of the parameters.

But there exist problems where the automata network approach supplies an exact, theoretically well understood, method.

These problems concern the optimization of norms in unitary balls and the computing of spectral norms. This last problem was studied by the school of numerical analysis represented by N. Gastinel, F. Robert and A. S. Householder [Ga,Ro,Ho]. The general problem consists in evaluating for a couple of norms φ, ψ in $\mathbb{R}^n$ the quantity:

$$S_{\varphi\psi} = \sup_{x\neq 0}\left(\frac{\varphi(x)}{\psi(x)}\right)$$

For instance, this quantity appears in the matrix spectral norms. In fact, is φ is a norm and A is a non singular matrix, then $\varphi \circ A(x) = \varphi(Ax)$ is also a norm. In this case

$$S_{\varphi\circ A,\varphi} = \sup_{x\neq 0}\left(\frac{\varphi(Ax)}{\varphi(x)}\right) = \|A\|_\varphi \text{ is a spectral norm of } A.$$

For instance, if $\varphi(x) = \|x\|_\infty$ then $\|A\|_\varphi = \max |a_{ij}|$, if $\varphi(x) = \|x\|$, then $\|A\|_\varphi = \sum_{i=1}^{n}\sum_{j=1}^{n} |a_{ij}|$ and if $\varphi(x) = \|x\|_2$ then $\|A\|_\varphi = \rho(A)$ the spectral radii of A.

For evaluating the optimum values $S_{\varphi\psi}$ or $\|A\|_\varphi$; F. Robert [Ro] has developed an iterative scheme which is very similar to a generalized automata network dynamics. In this pioneer work he used the notion of decomposable norms, which is due to N. Gastinel [Ga]. These are the norms verifying $\varphi(x) =< \gamma(x), x >$ for some function $\gamma : \mathbb{R}^n \to \mathbb{R}^n$. Now, for a couple of decomposable norms

$$\varphi(x) =< \gamma_\varphi(x), x >, \quad \psi(x) =< \gamma_\psi(x), x >$$

Robert [Ro] has studied the iteration: $\ell(t) = \gamma_\varphi(x(t))$, $x(t+1) = \gamma_\psi(\ell(t))$ for $t \geq 0$, starting from a point $x(0)$ such that $\psi^*(x(0)) = 1$, where ψ^* is the dual norm associated to ψ. The convergence study is made by showing that $\varphi(x(t)) \leq \psi(\ell(t)) \leq \varphi(x(t+1))$, i.e. by a Lyapunov functional technique. In the particular case $\varphi(x) = \psi(x) =< x, Bx >$ for some symmetric non-singular matrix, Robert's algorithm reduces to the well known power method to calculate the maximum eigenvalue of a symmetric matrix.

Robert's scheme is generalized in section 4 and we show that this generalization covers the dynamics of a $(\underline{P})$ automata network (introduced in [GM], we called them positive automata networks) in section 5. Moreover, we show that the functionals associated with the generalized Robert's algorithm corresponds to the usual Lyapunov functionals of $(\underline{P})$ automata networks.

Let us recall that in the $(\underline{P})$ automata networks the sign or threshold function which is the usual local function in neural networks, is replaced by a more general case, the $(\underline{P})$ functions in $\mathbb{R}^n$. The parallel iteration (1.3) of a $(\underline{P})$ automata network is

$$x(t+1) = \gamma(Ax(t) - b) \text{ for } t \geq 0 \tag{1.5}$$

where $\gamma : \mathbb{R}^n \to \mathbb{R}^n$ is a $(\underline{P})$ function. The neural network evolution corresponds to the case

$$\gamma(x) = \text{ sign } (x) = (\text{ sign } x(1), ..., \text{ sign } x(n)).$$

Hence, we have established a link between optimization of norms restricted to norm contraints and the evolution of neural networks.

Furthermore the duality of norm maximization problem, which appears in the work of Robert corresponding to the stationary points of its algorithm, can be extended in several directions. This program is fully developped in section 3, mainly in Theorems 3.2 and 3.3.

Finally, in section 5, we obtain as a corollary of these last duality theorems. This result was firstly obtained by Janssen, which reduces a class of discrete optimization problems to continuous problems [J].

2. Subnorms

In this section, we introduce the convex analysis framework. We relate $(\underline{P})$ functions and subnorms, in fact we show that the $(\underline{P})$ functions γ vanishing at 0, correspond to the decomposition of the subnorms $\varphi(x) =< \gamma(x), x >$. Reciprocally, we prove that any subnorm is decomposable in this way. We after supply some elementary results on dual subnorms. There are two simple results which are extremely useful tools: the Hölder inequality and the equality $\varphi(\gamma_\varphi(x)) = 1$ for any $\varphi(x) \neq 0$.

Let E be a linear space and $g : E \to I\!R \cup \{+\infty\}$ be an extended real function defined on E. The function g is called convex if

$$g(\lambda x + (1-\lambda)y) \leq \lambda g(x) + (1-\lambda)g(y) \quad \forall x, y \in E, \lambda \in [0,1].$$

By E' we mean the algebraic dual space: $E' = \{\ell : E \to I\!R \text{ linear }\}$. For $\ell \in E'$ and $x \in E$ we denote $< \ell, x >= \ell(x)$.

The functional $\gamma_g(x) \in E'$ is said to be a subgradient of g at x if:

$$g(y) \geq g(x) + < \gamma_g(x), y - x > \quad \text{for any } y \in E$$

The mapping $\gamma_g : E \to E', x \to \gamma_g(x)$, is a subgradient function of g if for any $x \in E$ the point $\gamma_g(x) \in E'$ is a subgradient of g at x.

It can be shown that a real function $g : E \to I\!R$ is convex if and only if it possesses subgradient function γ_g (the if part is direct, for the reciprocal see [Gi]).

If E is a locally convex topological linear space, denote by E^* its topological dual space: $E^* = \{\ell : E \to I\!R \text{ linear and continuous}\}$. Considering $g : E \to I\!R$ a continuous real function it can be proved that g is convex if and only if it has a subgradient function $\gamma_g : E \to E^*$, i.e. the subgradient $\gamma_g(x)$ is a continuous linear functional on E (see [L,R]).

A real function g defined on a linear space E is called

positive-homogeneous if $g(\lambda x) = \lambda g(x) \quad \forall x \in E,\ \lambda \in I\!R_+$;

subadditive if $g(x+y) \leq g(x) + g(y)\ \forall x, y \in E$.

If g is a positive-homogeneous function then it is subadditive if and only if it is convex.

Observe that any positive-homogeneous function vanishes at 0: $g(0) = 0$.

Our main results deal with subnorms. A real function $\varphi : E \to I\!R$ is a subnorm if it is positive-homogeneous, subadditive and non-negative (i.e. $\varphi(x) \geq 0$ for any $x \in E$). If φ verifies these properties but takes the value $+\infty$ we call it an extended subnorm. The subnorm φ is definite if $\varphi(x) = 0$ only when $x = 0$.

A seminorm φ is a symmetric subnorm, i.e. $\varphi(-x) = \varphi(x)$, so $\varphi(\lambda x) = |\lambda|\varphi(x)$ for any $\lambda \in I\!R$, $x \in E$. A definite seminorm is a norm. A symmetric extended subnorm is called extended seminorm, if it is also definite it is called extended norm.

From definition, subnorms and extended subnorms φ are convex functions and $\varphi(0) = 0$.

If E is a separated topological linear space a real convex function is continous if it is continuous at 0 [L]. If φ is a subnorm this is equivalent to the fact that the set $\{x \in E : \varphi(x) \leq 1\}$ is a neighbourhood at 0 [S].

In $E = I\!R^n$ any real convex function is continuous, so any subnorm in $I\!R^n$ is continuous.

In order to characterize the subgradient functions of positive-homogeneous convex functions we introduce the following useful definition.

Definition 2.1. [GM] $\gamma : E \to E'$ is said to be a $(\underline{P})$ function if

$$< \gamma(x) - \gamma(y), x > \geq 0 \quad \text{for any } x, y \in E. \quad \blacksquare$$

Now, associate to any $\gamma : E \to E'$ the real function

$$\varphi_\gamma(x) = < \gamma(x), x > \text{ for } x \in E. \tag{2.1}$$

We have,

Lemma 2.1. $\gamma : E \to E'$ is a $(\underline{P})$ function if and only if the function φ_γ is convex and γ is a subgradient function of it.

Proof. It follows from the equivalence:

$$\varphi_\gamma(x) \geq \varphi_\gamma(y) + < \gamma(y), x - y > \Longleftrightarrow < \gamma(x) - \gamma(y), x > \geq 0 \quad \blacksquare$$

The main limits between $(\underline{P})$ functions, positive homogeneous convex functions and seminorms are supplied in the following two propositions, which in a less complete form, were shown in [GM].

Proposition 2.1. Let E be a linear space and φ be a convex function defined on E. Then, the following two conditions, are equivalent:

(i) φ is positive-homogeneous.
(ii) Any subgradient γ_φ of φ is a $(\underline{P})$ function.

When this condition holds we have that:

$$\varphi = \varphi_\gamma \text{ for any subgradient function } \gamma \text{ of } \varphi, \tag{2.2}$$

and there exists a subgradient γ_φ of φ such that:

$$\gamma_\varphi(\lambda x) = \gamma_\varphi(x) \text{ for any } \lambda > 0 \text{ and } x \in E \tag{2.3}$$

Proof. (i) $\Longrightarrow$ (ii) If φ is a positive-homogeneous subadditive function then it is convex. We denote by γ_φ a subgradient of it. We have:

$$2\varphi(x) = \varphi(2x) \geq \varphi(x) + < \gamma_\varphi(x), x >$$

On the other hand φ positive-homogeneous implies $\varphi(0) = 0$. Therefore

$$0 = \varphi(0) \geq \varphi(x) + < \gamma_\varphi(x), -x >$$

We deduce $\varphi(x) = \varphi_{\gamma_\varphi}(x) =< \gamma(x), x >$. The convexity of φ_{γ_φ} implies that γ_φ verifies property $(\underline{P})$. In particular we have shown that (i) implies equality (2.2).

(ii) $\Longrightarrow$ (i) Since γ_φ is a $(\underline{P})$ function, $\varphi = \varphi_{\gamma_\varphi}$ is convex. Let us show that it is positive homogeneous.

From $< \gamma_\varphi(\lambda x) = \gamma_\varphi(x), \lambda x >\geq 0$ and $< \gamma_\varphi(x) - \gamma_\varphi(\lambda x), x >\geq 0$ for any $\lambda > 0$, $x \in E$ we deduce $< \gamma_\varphi(\lambda x) - \gamma_\varphi(x), x >= 0$. So, $\varphi(\lambda x) = \lambda \varphi(x)$ for any $\lambda \geq 0$, $x \in E$.

Now let us show (2.3). The relation $x \sim y$ if and only if $y = \lambda x$ for some $\lambda > 0$ is an equivalence relation. A ray L is an equivalence class of $\sim$. Choose a unique point x_L for each L. Let γ be a subgradient function of φ. Define γ' by $\gamma'(x) = \gamma(x_L)$ when $x \in L$. We shall prove γ' is also a subgradient function of φ.

Take $x, y \in E$. We have $x = \lambda x_L$ for some x_L and $\lambda > 0$ so $\varphi(x) = \lambda \varphi(x_L)$. Now

$$\begin{aligned}\varphi(y) = \lambda\varphi(\lambda^{-1}y) &\geq \lambda(\varphi(x_L) + < \gamma(x_L), \lambda^{-1}y - x_L >) \\ &= \varphi(x) + < \gamma(x'), y - x >\end{aligned}$$

Then $\gamma_\varphi = \gamma'$ is a subgradient of φ and verifies the condition (2.3). ■

Proposition 2.2. Let E be a linear space and φ be a convex function defined on E. Then they are equivalent:

(i) φ is a subnorm.
(ii) There exists a subgradient function γ_φ of φ which is a $(\underline{P})$ function and $\gamma_\varphi(0) = 0$.

When this condition holds we can choose a subgradient function γ_φ of φ verifying condition (2.3) and such that

$$\gamma_\varphi(x) \neq 0 \text{ if and only if } \varphi(x) > 0. \tag{2.4}$$

In particular, this last means that φ is definite if $\gamma_\varphi(x) = 0$ is equivalent to $x = 0$.

Moreover, φ is a seminorm if and only if there exists a subgradient function γ_φ which is a $(\underline{P})$ and antisymmetric function, where antisymmetric means: $\gamma_\varphi(-x) = -\gamma_\varphi(x)$ for any $x \in E$. Finally, γ is a norm if and only if there exists a subgradient function γ_φ which is a $(\underline{P})$ and antisymmetric function which only vanishes at 0.

Proof. (i) $\Longrightarrow$ (ii) φ attains its minimum at $0 \in E$ so we can choose $\gamma_\varphi(0) = 0$. From Proposition 1, γ_φ is a $(\underline{P})$ function.

(ii) $\Longrightarrow$ (i) From Proposition 2.1, φ is positive-homogeneous so $\varphi(0) = 0$. Being $\gamma_\varphi(0) = 0$ we deduce φ attains its minimum at $0 \in E$, so $\varphi(x) \geq \varphi(0) = 0$ $\forall x \in E$.

On the other hand for any other minimum $y \in E$ we can choose a subgradient function γ of φ verifying $\gamma(y) = 0$. For this subgradient we define $\gamma' = \gamma_\varphi$ as we made in the proof of condition (2.3), so this subgradient function verifies both conditions (2.3) and (2.4). So, when φ is definite we have $\gamma_\varphi(x) = 0$ only in the case $x = 0$.

Now, let φ be a seminorm and γ a subgradient of it. Then $< \gamma(x), x >= \varphi(x) = \varphi(-x) =< \gamma(-x), -x >$. Consider the antisymmetric function $\gamma'(x) = \frac{\gamma(x)-\gamma(-x)}{2}$, then $\varphi(x) =< \gamma'(x), x >$. Let us show γ' is a $(\underline{P})$ function. We have

$$< \gamma'(x) - \gamma'(y), x >= \frac{1}{2}(< \gamma(x) - \gamma(y), x > + < \gamma(-x) - \gamma(-y), -x >)$$

which is ≥ 0. Then $\gamma_\varphi = \gamma'$ is a $(\underline{P})$ and antisymmetric subgradient of φ. The reciprocal is evident, if γ_φ is a $(\underline{P})$ and antisymmetric function then φ is a seminorm. From our results it follows that φ is a norm (i.e. a definite seminorm) if there exists a subgradient function γ_φ which is a $(\underline{P})$ and antisymmetric function vanishing only at $x = 0$. ■

For a subnorm φ we only consider the subgradient functions γ_φ verifying the conditions set in Propositions 2.1 and 2.2.

Definition 2.2. Let E be a linear space and $\varphi : E \to I\!R \cup \{+\infty\}$ be an extended real function. Define $\varphi' : E' \to I\!R \cup \{+\infty\}$ by:

$$\varphi'(\ell) = \sup\{< \ell, x >: x \in E \text{ such that } \varphi(x) \leq 1\} \tag{2.5}$$

where $\sup \phi = 0$. ■

Proposition 2.3. φ' is an extended subnorm on E'. If φ is symmetric, i.e. $\varphi(-x) = \varphi(x)$ for any $x \in E$, then φ' is an extended seminorm on E'. Moreover, if φ is a norm then φ' is an extended norm.

Proof. By definition $\varphi'(\ell) \geq 0$. On the other hand $\varphi'(0) = 0$ and φ' is positive homogeneous.

Let $h^x \in E'' = (E')'$ be the following functional on E': $h^x(\ell) =< \ell, x >$ which is linear, so convex. We have $\varphi' = \sup\{h^x : x \in E, \varphi(x) \leq 1\}$. Since the sup of convex functions is convex we deduce the convexity of φ'. So φ' is an extended subnorm on E'.

Let us assume φ symmetric. From $< -\ell, x > = < \ell, -x >$ we deduce that $\varphi'(-\ell) = \varphi'(\ell)$ so φ' is an extended seminorm. Now let φ be a norm, and $\varphi'(\ell) = 0$. Then $< \ell, x > \leq 0$ for any $x \in E$. From $< \ell, x > \leq 0$ and $< \ell, -x > \leq 0$ we get $< \ell, x > = 0$ for any $x \in E$, so $\ell = 0$. ■

Let φ be a subnorm on E, then φ' is called the dual extended subnorm of φ. In this case the following relation which is a generalized Hölder inequality, is verified:

$$< \ell, x > \leq \varphi(x) \cdot \varphi'(\ell) \text{ when } \varphi(x) > 0 \tag{2.6}$$

Proposition 2.4. Let γ_φ be a subgradient function of the subnorm φ on E. Then:

$$\varphi'(\gamma_\varphi(x)) = 1 \quad \text{for any } \gamma_\varphi(x) \neq 0 \tag{2.7}$$

Proof. If $\gamma_\varphi(x) \neq 0$ then $\varphi(x) > 0$. So, by (2.6):

$$< \gamma_\varphi(x), x > \leq \varphi(x) \cdot \varphi'(\gamma_\varphi(x))$$

Since $\varphi(x) = < \gamma_\varphi(x), x >$ we deduce $\varphi'(\gamma_x) \geq 1$.

For proving the another inequality take $y \in E$ such that $\varphi(y) = 1$. By using property $(\underline{P})$ of γ we get: $< \gamma_\varphi(x), y > \leq < \gamma_\varphi(y), y > = \varphi(y) = 1$. So $\varphi'(\gamma_\varphi(x)) \leq 1$. ■

Let φ be a subnorm, γ be a subgradient function of it and $x \in E$ be such that $\varphi(x) > 0$. Then $\frac{1}{\varphi(x)} h^x \in E''$, is a subgradient of φ' at point $\gamma_\varphi(x)$. In fact, from (2.6) and (2.7) we obtain:

$$\varphi'(\gamma_\varphi(x)) + < \frac{1}{\varphi(x)} h^x, \ell - \gamma_\varphi(x) > = < \ell, \frac{x}{\varphi(x)} > \leq \varphi'(\ell)$$

Now, for any extended real function φ defined on E we denote:

$$\ker \varphi = \{x \in E : \varphi(x) = 0\} \tag{2.8}$$

If φ is a subnorm the set $\ker \varphi$ is a cone on E : $x, y \in \ker \varphi$ imply $\lambda x + \mu y \in \ker \varphi$ for any $\lambda, \mu \geq 0$.

Now let us consider the extended subnorm $\varphi'' = (\varphi')'$ defined on the linear space $E'' = (E')'$ (recall that this space contains E). We have:

Proposition 2.5. Let φ be a subnorm on E. Then, the extended subnorm φ'' defined on E'' verifies:

(i) $\ker \varphi'' \cap E \subset \ker \varphi$.

(ii) $\varphi''|_{E-\ker\varphi} = \varphi|_{E-\ker\varphi}$
(iii) $\varphi''|_E = \varphi$ if and only if $\ker\varphi \subset \ker\varphi'' \cap E$.

Proof. Let $x \in E$ be such that $\varphi(x) > 0$.

We have $\varphi''(x) = \sup\{< \ell, x >: \ell \in E'$ such that $\varphi'(\ell) \leq 1\}$. Since $\varphi'(\gamma_\varphi(x)) = 1$ we have: $\varphi''(x) \geq < \gamma_\varphi(x), x > = \varphi(x) > 0$. Then (i).

Now, since: $< \ell, x > \leq \varphi(x) \cdot \varphi'(\ell)$, we deduce $\varphi''(x) \leq \varphi(x)$. Therefore (ii): $\varphi''(x) = \varphi(x)$ for any $x \in E$ such that $\varphi(x) > 0$.

Last property follows from (i), (ii). ■

Then, for a subnorm φ on E, the function $\tilde{\varphi} = \varphi''|_E$ is an extended subnorm verifying $\ker\tilde{\varphi} \subset \ker\varphi$ and $\tilde{\varphi}|_{E-\ker\varphi} = \varphi|_{E-\ker\varphi}$. Moreover, it also satisfied the bidual equality $\tilde{\varphi}''|_E = \tilde{\varphi}$. In fact, from Proposition 2.5 (iii) it results that the bidual equality holds if $\ker\tilde{\varphi} \subset \ker\tilde{\varphi}'' \cap E$. Now, if $\tilde{\varphi}(x) = 0$ then $< \ell, x > \leq 0$ for any $\ell \in E'$, so $\tilde{\varphi}''(x) = \sup\{< \ell, x >: \ell \in E', \tilde{\varphi}'(\ell) \leq 1\} \leq 0$.

When E is a topological vector space and φ is an extended real function we can define φ' on the topological dual space E^* by the same formula (2.5). We denote it by $\varphi^* = \varphi'|_{E^*}$. Since $<,>: E^* \times E \to \mathbb{R}$ is continuous in both coordinates and

$$\varphi^*(\cdot) = \sup\{< x, \cdot >: \varphi(x) \leq 1\}$$

we deduce that φ^*, defined on E^*, is lower semi-continuous.

When φ is a continuous subnorm then its subgradient function verifies $\gamma_\varphi(x) \in E^*$ for any $x \in E$, so formula (2.7): $\varphi^*(\gamma_\varphi(x)) = 1$ for any $\gamma_\varphi(x) \neq 0$, also holds for the dual extended subnorm φ^* defined on E^*. Moreover if φ is a continuous seminorm then φ^* is an extended seminorm on E^* and if φ is a continuous norm then φ^* is an extended norm on E^*.

If φ is a norm generating the topology of E then φ^* is a norm on E^*. This is the case for any norm defined on $\mathbb{R}^n$.

The extended subnorm φ^{**} is defined on E^{**}, the topological dual space of E^*. Properties (i), (ii), (iii) of Proposition 2.5 are verified when we replace φ'' by φ^*. Then $\tilde{\varphi} = \varphi^{**}|_E$ is an extended subnorm verifying $\ker\tilde{\varphi} \subset \ker\varphi$, $\tilde{\varphi}|_{E-\ker\varphi} = \varphi|_{E-\ker\varphi}$ and the bidual equality $\tilde{\varphi}^{**}|_E = \tilde{\varphi}$. If E and E^* are in duality, i.e. $E^{**} = E$, and φ defines the topology of E, then $\varphi^{**} = \varphi$.

3. Duality of Max Subnorm Problems

In this section we supply the duality between two subnorm maximization problems restricted to the unit "ball" of the extended dual subnorms. To establish our results we are led to impose some kernel equalities which become trivial when we deal with norms. Theorem 3.1 is esentially a generalization of a Robert's result. In Theorem 3.2 and 3.2 we give a framework for the duality of maximum of subnorms. This new frame not only covers the Robert's result but also a recent result established by Janssen [J], which we develop in section 5.

Let E be a linear space and $F \subset E'$ be a linear subspace of the algebraic dual space. Let φ and ψ be subnorms defined on E and F respectively, such that $\gamma_\varphi(x) \in F$ for any $x \in E$ and $\gamma_\varphi(\ell) \in E$ for any $\ell \in F$.

Let us observe that this situation occurs if E and F are locally convex topological linear spaces in duality, i.e. $F = E^*$ and $E = F^*$, and φ and ψ are continuous subnorms on E and F respectively.

By φ' and ψ' we shall also mean the restrictions of the dual extended subnorms to F and E i.e. $\varphi' = \varphi'|_F$, $\psi' = \psi'|_E$. In the topological case we denote the dual extended subnorms by φ^*, ψ^*.

By γ_φ and γ_ψ, we denote some subgradient functions of φ and ψ respectively. We have $\gamma_\varphi(x) \in F$ for any $x \in E$ and $\gamma_\psi(\ell) \in E$ for any $\ell \in F$.

The following inclusions are verified:

$$\ker \varphi' \subseteq \ker(\varphi \circ \gamma_\psi) \text{ and } \ker \psi' \subseteq \ker(\psi \circ \gamma_\varphi). \tag{3.1}$$

In fact, the equality $\varphi'(\ell) = 0$ implies $< \ell, x > \leq 0$ for any $x \notin \ker \varphi - \{0\}$. Since $\psi(\ell) =< \ell, \gamma_\psi(\ell) >$ we deduce $\psi(\ell) = 0$ or $\gamma_\psi(\ell) \in \ker \varphi - \{0\}$, so $\varphi(\gamma_\psi(\ell)) = 0$. The second inclusion is shown in an analogous way.

Observe that for any $x \in E$ with $\gamma_\varphi(x) \neq 0$ (i.e. $\varphi(x) > 0$) the Proposition 7 implies that the point $\gamma_\varphi(x) \in F$ verifies $\varphi'(\gamma_\varphi(x)) = 1$. Similarly for any $\ell \in F$ with $\gamma_\psi(\ell) \neq 0$ (i.e. $\psi(\ell) > 0$) the point $\gamma_\varphi(\ell) \in E$ verifies $\psi'(\gamma_\psi(\ell)) = 1$.

Lemma 3.1. Let us assume $\ker \varphi' = \ker(\varphi \circ \gamma_\psi)$, $\ker \psi' = \ker(\psi \circ \gamma_\varphi)$. Then:

(i) If $\psi'(x) = 1$ then $\varphi'(\gamma_\varphi(x)) = 1$ and $\varphi(x) \leq \psi(\gamma_\varphi(x))$
(ii) If $\varphi'(\ell) = 1$ then $\psi'(\gamma_\psi(\ell)) = 1$ and $\psi(\ell) \leq \varphi(\gamma_\psi(\ell))$

Proof. (i) From hypothesis we get $\psi(\gamma_\varphi(x)) > 0$. This implies $\varphi'(\gamma_\varphi(x)) = 1$. We can apply inequality (2.6) so as to obtain:

$$\varphi(x) =< \gamma_\varphi(x), x > \leq \psi(\gamma_\varphi(x))\psi'(x) = \psi(\gamma_\varphi(x)).$$

Property (ii) is analogously shown. ■

Remark. Observe that the hypotheses set in the last Lemma:

$$\ker \varphi' = \ker(\varphi \circ \gamma_\psi), \ \ker \psi' = \ker(\psi \circ \gamma_\varphi)$$

are weaker than the following ones:

$$\ker \varphi' = \ker \psi, \ \ker \psi' = \ker \varphi$$

Let us show it. We assume then that the last two equalities are verified. From (3.1) it suffices to prove that $\ker(\varphi \circ \gamma_\psi) \subseteq \ker \psi'$. Now if $\varphi(\gamma_\psi(x)) = 0$ then $\gamma_\psi(x) \in \ker \varphi = \ker \psi'$ so $\psi'(\gamma_\psi(x)) = 0$. From Proposition 2.4 we get $\gamma_\psi(x) = 0$, so $\psi(x) = 0$. Then $\varphi'(x) = 0$. We have shown the first inclusion, the another one is proved in a similar way. ■

Now we are able to generalize a duality found by Robert for norms in $\mathbb{R}^n$.

Theorem 3.1. Let E and $F = E^*$ be locally convex topological vector spaces in duality and φ and ψ be continuous subnorms on E and F respectively. Assume $\ker \varphi^* = \ker(\varphi \circ \gamma_\psi)$, $\ker \psi^* = \ker(\psi \circ \gamma_\varphi)$. Then:

$$\sup_{x \in E : \psi^*(x) = 1} \varphi(x) = \sup_{\ell \in F : \varphi^*(\ell) = 1} \psi(\ell) \tag{3.2}$$

Proof. Let $x \in E$ be such that $\psi^*(x) = 1$. Take $\ell = \gamma_\varphi(x)$. From Lemma 3.1 (i): $\psi^*(\ell) = 1$ and $\varphi(x) \leq \psi(\ell)$. Then the inequality $\leq$ holds in (3.2). From Lemma 3.1 (ii) we get $\geq$. ■

When φ and ψ are norms the kernel hypotheses are verified, so:

Corollary 3.1. Let E and $F = E^*$ be locally convex topological vector spaces in duality and φ and ψ be continuous norms on E and F respectively. Then the equality (3.2) holds. ■

Let us extend these last results. Consider $B : E \to E$ a linear mapping, by $B' : E' \to E'$ we denote the dual mapping $(B'\ell)(x) = \ell(Bx)$.

Lemma 3.2. Assume $\ker \varphi' = (B')^{-1}(\ker \psi)$, $\ker \psi' = B^{-1}(\ker \varphi)$ and $B'F \subset F$. Then:

(i) If $\psi'(x) = 1$ then $\varphi'(\gamma_\varphi(Bx)) = 1$ and $\varphi(Bx) \leq \psi(B'\gamma_\varphi(Bx))$
(ii) If $\varphi'(\ell) = 1$ then $\psi'(\gamma_\psi(B'\ell)) = 1$ and $\psi(B'\ell) \leq \varphi(B\gamma_\psi(B'\ell))$

Proof. (i) From hypotheses $\psi'(x) = 1$ implies $\varphi(Bx) > 0$. Then $\gamma_\varphi(Bx) \neq 0$ and $\varphi'(\gamma_\varphi(Bx)) = 1$. From hypotheses we deduce $\psi(B'\gamma_\varphi(Bx)) > 0$. On the other hand, $\varphi(Bx) = \langle \gamma_\varphi(Bx), Bx \rangle = \langle B'\gamma_\varphi(Bx), x \rangle$. Hence we can apply inequality (2.6) to obtain:

$$\varphi(Bx) \leq \psi(B'\gamma_\varphi(Bx))\psi'(x) = \psi(B'\gamma_\varphi(Bx))$$

Property (ii) is proved in a similar way. ■

Proposition 3.1. Let φ and ψ be subnorms on E and F respectively. Assume $\ker\varphi' = (B')^{-1}(\ker\psi)$, $\ker\psi' = B^{-1}(\ker\varphi)$. Then for any linear mapping $B : E \to E$ such that $B'F \subset F$ we have:

$$\sup_{x\in E:\psi'(x)=1} \varphi(Bx) = \sup_{\ell\in F:\varphi'(\ell)=1} \psi(B'\ell) \tag{3.3}$$

Proof. Take $x \in E$ such that $\psi'(x) = 1$. Let us denote $\ell = \gamma_\varphi(Bx)$. From Lemma 3.2 (i) we get $\varphi'(\ell) = 1$ and $\varphi(Bx) \leq \psi(B'\ell)$. Then inequality $\leq$ holds in expression (3.3). Inequality $\geq$ follows from Lemma 3.2 (ii). ■

As a corollary we obtain the following result for the topological case:

Theorem 3.2. Let E and $F = E^*$ be locally convex topological linear spaces in duality. Let φ and ψ be continuous subnorms on E and F respectively, and $B : R \to E$ be a continuous linear operator. Assume $\ker\varphi^* = (B^*)^{-1}(\ker\psi)$, $\ker\psi^* = B^{-1}(\ker\varphi)$. Then,

$$\sup_{x\in E:\psi^*(x)=1} \varphi(Bx) = \sup_{\ell\in E^*:\varphi^*(\ell)=1} \psi(B^*\ell) \tag{3.4}$$

Proof. $F = E^*$ and $B^*E^* \subset E^*$, so our last result can be applied. ■

When φ and ψ are norms the kernel conditions are verified. We get,

Corollary 3.2 Let E and $F = E^*$ be locally convex topological linear spaces in duality, φ and ψ be continuous norms on E and F respectively, and $B : E \to E$ be a linear operator. Then, equality (3.4) holds. ■

We shall give a new insight into these last results in the special case of Hilbert spaces. Let φ and ψ be respectively continuous subnorms on the Hilbert spaces H and $\mathcal{H}$ respectively. Since we can identify $H = H^*$ and $\mathcal{H} = \mathcal{H}^*$ the extended subnorms φ^* and ψ^* are also defined on H and $\mathcal{H}$ respectively. Let $B : H \to \mathcal{H}$ be a continuous operator and

$$B^* : \mathcal{H} \to H \text{ be its adjoint operator: } < B^*\ell, x > = < \ell, Bx > .$$

Lemma 3.3. Assume $\ker\varphi^* = B^{-1}(\ker\psi)$ and $\ker\psi^* = (B^*)^{-1}(\ker\varphi)$. Then:

(i) If $\psi^*(\ell) = 1$ then $\varphi^*(\gamma_\varphi(B^*\ell)) = 1$ and $\varphi(B^*\ell) \leq \psi(B\gamma_\varphi(B^*\ell))$
(ii) If $\varphi^*(x) = 1$ then $\psi^*(\gamma_\psi(Bx)) = 1$ and $\psi(Bx) \leq \varphi(B^*\gamma_\psi(Bx))$

Proof. (i) From hypotheses $\psi^*(\ell) = 1$ implies $\varphi(B^*\ell) > 0$. Then:

$$\varphi(B^*\ell) = < B^*\ell, \gamma_\varphi(B^*\ell) > = < B\gamma_\varphi(B^*\ell), \ell > > 0$$

Hence $\gamma_\varphi(B^*\ell) \neq 0$ and $\varphi^*(\gamma_\varphi(B^*\ell)) = 1$. From hypotheses $\psi(B\gamma_\varphi(B^*\ell)) > 0$ then we can apply inequality (2.6) to obtain:

$$\varphi(B^*\ell) \leq \psi(B\gamma_\varphi(B^*\ell))\psi^*(\ell) = \psi(B\gamma_\varphi(B^*\ell))$$

Property (ii) is shown in an analogous way. ■

Theorem 3.3. Let H and $\mathcal{H}$ be Hilbert spaces. Let φ and ψ be continuous subnorms on H and $\mathcal{H}$ respectively and $B : H \to \mathcal{H}$ be a continuous linear operator. Assume $\ker \varphi^* = B^{-1}(\ker \psi)$, $\ker \psi^* = (B^*)^{-1}(\ker \varphi)$. Then,

$$\sup_{\ell \in \mathcal{H}: \psi^*(\ell)=1} \varphi(B^*\ell) = \sup_{x \in H: \varphi^*(x)=1} \psi(Bx) \tag{3.5}$$

Proof. $\psi^*(\ell) = 1$ we take $x = \gamma_\varphi(B^*\ell)$ which satisfies $\varphi^*(x) = 1$ by Lemmas 3.3 (i). Moreover $\varphi(B^*\ell) \leq \psi(Bx)$. Then inequality $\leq$ holds in (3.5). Analogously, from Lemma 3.3 (ii) we show that $\geq$ is also satisfied. ■

We deduce,

Corollary 3.3. Let H and $\mathcal{H}$ be Hilbert spaces, φ and ψ be continuous norms on H and $\mathcal{H}$ respectively and $B : H \to \mathcal{H}$ be a continuous linear operator. Then equality (3.5) is verified. ■

4. Robert's Scheme and Its Generalization

In this section we develop the framework established by Robert [Ro] for the study of eigenvectors and eigenvalues of couple of norms in $\mathbb{R}^n$. This framework includes the definitions, an algorithm for the eigenvectors and eigenfunctions computation as well as the proof that this algorithm contains the usual power method. We summarize the key results of Robert in Theorems 4.1 and 4.2. Robert's scheme is generalized at the end of this section. In fact, we deal with norms φ and ψ defined on spaces $\mathbb{R}^n$ and $\mathbb{R}^m$ and with a general $m \times n$ real matrix B. In this context the results of Robert are reobtained by taking $m = n$ and B the identity matrix in our Theorem 4.3.

Definition 4.1. [Ro] Let φ_1, φ_2 be subnorms defined on the locally convex topological linear space E. A vector $\bar{x} \in E$ is said to be an eigenvector of the subnorms (φ_1, φ_2), we denote $\bar{x} \in P(\varphi_1, \varphi_2)$, if there exists $\bar{\ell} \in E^*$ such that

$$< \bar{\ell}, \bar{x} >= \varphi_1(\bar{x})\varphi_1^*(\bar{\ell}) = \varphi_2(\bar{x})\varphi_2^*(\bar{\ell}) \tag{4.1}$$

The value $\lambda(\bar{x}) = \frac{\varphi_1(\bar{x})}{\varphi_2(\bar{x})}$ is called the eigenvalue associated to $\bar{x} \in P(\varphi_1, \varphi_2)$, we denote $\lambda(\bar{x}) \in \Lambda(\varphi_1, \varphi_2)$. ■

Remark that if $\bar{x} \in P(\varphi_1, \varphi_2)$ and $\bar{\ell} \in E^*$ verifies (4.1) then $\bar{\ell} \in P(\varphi_2^*, \varphi_1^*)$ and:

$$\lambda(\bar{\ell}) = \frac{\varphi_2^*(\bar{\ell})}{\varphi_1^*(\bar{\ell})} = \frac{\varphi_1(\bar{\ell})}{\varphi_2(\bar{x})} = \lambda(\bar{x}) \tag{4.2}$$

On the other hand, if $\bar{x}$ is also considered as an element of $P(\varphi_2, \varphi_1)$ then its eigenvalue is the reciprocal of $\lambda(\bar{x})$.

Now, let E and $F = E^*$ be locally convex topological linear spaces in duality. We consider that φ and ψ are continuous subnorms on E and F respectively. We denote by φ^* and ψ^* their dual (extended) subnorms on F and E respectively, and by γ_φ and γ_ψ the subgradient functions of φ and ψ. Assume:

$$\ker \varphi^* = \ker(\varphi \circ \gamma_\psi), \quad \ker \psi^* = \ker(\psi \circ \gamma_\varphi) \tag{4.3}$$

Let $x(0) \in E$ be such that $\psi^*(x(0)) = 1$. By induction define:

$$\ell(t) = \gamma_\varphi(x(t)), \quad x(t+1) = \gamma_\psi(\ell(t)) \text{ for } t \geq 0 \tag{4.4}$$

From Lemma 3.1 we deduce the following relations,

$$\psi^*(x(t)) = 1, \quad \varphi^*(\ell(t)) = 1, \quad \forall t \geq 0. \tag{4.5}$$

and,

$$\varphi(x(t)) \leq \psi(\ell(t)) \leq \varphi(x(t+1)) \quad \forall t \geq 0 \tag{4.6}$$

Then $(\varphi(x(t)))$ and $(\psi(\ell(t)))$ are increasing sequences and their limits are equal. We denote them by $\alpha(x(0))$,

$$\lim_{t\to\infty} \varphi(x(t)) = \lim_{t\to\infty} \psi(\ell(t)) = \alpha(x(0)), \tag{4.7}$$

because it only depends on the initial point $x(0)$.

We recall that the same result is obtained if we begin with $\ell(0) \in F$ such that $\varphi^*(\ell(0)) = 1$ and we take $x(t) = \gamma_\psi(\ell(t))$, $\ell(t+1) = \gamma_\varphi(x(t))$, for $t \geq 0$.

Now, we take $E = \mathbb{R}^n$, which is identified with $E^* = \mathbb{R}^{n*}$. We assume that φ and ψ are norms on $\mathbb{R}^n$, so they are continuous. The functions φ^* and ψ^* are also norms on $\mathbb{R}^n$. In this case the sets

$$S_{\psi^*} = \{x \in \mathbb{R}^n : \psi^*(x) = 1\} \text{ and } S_{\varphi^*} = \{\ell \in \mathbb{R}^n : \psi^*(\ell) = 1\}$$

are compact sets. Observe that in this case the quantity $\alpha(x(0))$ of (4.7) is finite because it is bounded by $\sup\{\varphi(x) : x \in S_{\psi^*}\}$, which is finite because φ is continuous and S_{ψ^*} is compact. The sequence $(x(t))$ lies in S_{ψ^*} then there exists a convergent subsequence $(x(t_k))$. Its limit is denoted by $\bar{x}$. Since the sequence $(\ell(t_k))$ is included in S_{φ^*}, it contains a convergent subsequence $(\ell(t_r))$, its limit is denoted by $\bar{\ell}$. From (4.5) and the continuity we obtain $\psi^*(\bar{x}) = \varphi^*(\bar{\ell}) = 1$.

From continuity and (4.7) we get

$$\alpha = \varphi(\bar{x}) = \psi(\bar{\ell}) = \lim_{t\to\infty} \varphi(x(t)) = \lim_{t\to\infty} \psi(\ell(t)).$$

Now

$$\varphi(x(t_r)) = < \gamma_\varphi(x(t_r)), x(t_r) > = < \ell(t_r), x(t_r) > .$$

Then $\alpha = < \bar{\ell}, \bar{x} >$. Then the generalized Hölder inequalities (2.6) for φ, ψ become equalities for $\bar{\ell}$, $\bar{x}$ in the sense that:

$$< \bar{\ell}, \bar{x} > = \varphi(\bar{x})\varphi^*(\bar{\ell}) = \psi(\bar{\ell})\psi^*(\bar{x}) \tag{4.8}$$

and

$$\frac{\varphi(\bar{x})}{\psi^*(\bar{x})} = \frac{\varphi^*(\bar{\ell})}{\psi(\bar{\ell})} = \alpha(x(0)) \tag{4.9}$$

We have obtained:

Theorem 4.1. [Ro] For any pair of norms φ, ψ on $\mathbb{R}^n$ let us consider the sequences defined by $\ell(t) = \gamma_\varphi(x(t))$, $x(t+1) = \gamma_\psi(\ell(t))$, for $t \geq 0$, with $\psi^*(x(0)) = 1$. Then $\lim_{t\to\infty} \varphi(x(t)) = \lim_{t\to\infty} \psi(\ell(t))$ and it is finite. Moreover, if we call $\alpha(x(0))$ this limit, there exists convergent subsequences $x(t_r) \to \bar{x}$, $\ell(t_r) \to \bar{\ell}$ such that $< \bar{\ell}, \bar{x} > = \alpha(x(0))$. We have $\bar{x} \in P(\varphi, \psi^*)$, $\bar{\ell} \in P(\psi, \varphi^*)$ and $\alpha(x(0))$ is the eigenvalue associated to $\bar{x}$. ■

Following Robert we develop the special case $\psi = \varphi$. In this case $x(t+1) = \gamma_\varphi(\ell(t))$, $\ell(t) = \gamma_\varphi(x(t))$. Then if we denote $y(2t) = x(t)$, $y(2t+1) = \ell(t)$ for $t \geq 0$ the following equation is verified:

$$y(t+1) = \gamma_\varphi(y(t)) \tag{4.10}$$

Let us recall that for any initial condition $y(0) \neq 0$ we get $\varphi^*(y(t)) = 1$ for any $t \geq 1$.

From (4.6), $\varphi(y(t))$ increases for $t \geq 1$. The set $\{y : \psi^*(y) = 1\}$ is compact so $\varphi(y(t))$ increases to a finite value $\alpha(y(0))$.

Denote by $\| \ \|$ the euclidean norm on $I\!R^n$. We have:
$\varphi(y(t)) =< y(t+1), y(t) >\leq \|y(t+1)\| \|y(t)\|$, so $\frac{\varphi(y(t))}{\|y(t+1)\| \|y(t)\|} \leq 1$.

On the other hand, $\varphi(y(t)) = \varphi^{**}(y(t)) \geq < \gamma_\varphi(y(t-1)), y(t) >$ because $\varphi^*(\gamma_\varphi(y(t-1))) = 1$. Then $\varphi(y(t)) \geq < y(t), y(t) >= \|y(t)\|^2$. Therefore:

$$\|y(t)\|^2 \leq \varphi(y(t)) \leq \|y(t+1)\| \|y(t)\| \tag{4.11}$$

This also implies $\|y(t)\|$ is increasing with $t \geq 1$. Hence, if we denote $\alpha(y(0)) = \lim_{t \to \infty} \varphi(y(t))$ we get $\alpha(y(0)) = \lim_{t \to \infty} \|y(t)\|^2$. Therefore:

$$\begin{aligned} \|y(t+1) - y(t)\|^2 &= \|y(t+1)\|^2 + \|y(t)\|^2 - 2 < y(t+1), y(t) > \\ &= \|y(t+1)\|^2 + \|y(t)\|^2 - 2\varphi(y(t)) \underset{t \to \infty}{\longrightarrow} 0. \end{aligned}$$

The sequence $(y(t))_{t \geq 1}$ belongs to the compact set S_{φ^*} so there exists a converging subsequence $y(t_k) \underset{k \to \infty}{\longrightarrow} \bar{y}$. Since $\|y(t_k+1) - y(t_k)\| \underset{k \to \infty}{\longrightarrow} 0$ we also deduce $y(t_k + 1) \underset{k \to \infty}{\longrightarrow} \bar{y}$. From continuity:

$$\begin{aligned} &\varphi(y(t_k)) \underset{k \to \infty}{\longrightarrow} \varphi(\bar{y}) = \alpha(y(0)), \quad \varphi(y(t_k)) =< y(t_k + 1), y(t_k) > \underset{k \to \infty}{\longrightarrow} < \bar{y}, \bar{y} > \\ &\|y(t_k)\|^2 \underset{k \to \infty}{\longrightarrow} \|\bar{y}\|^2 = \alpha(y(0)), \quad \varphi^*(y(t_k)) \underset{k \to \infty}{\longrightarrow} \varphi^*(\bar{y}) = 1 \end{aligned} \tag{4.12}$$

Then the limit point verifies:

$$\varphi^*(\bar{y}) = 1, < \bar{y}, \bar{y} >= \varphi(\bar{y}) = \alpha(y(0))$$

Hence:

Theorem 4.2. [Ro] For any norm φ on $I\!R^n$ the sequence $y(t) = \gamma_\varphi(y(t-1))$, with $y(0) \neq 0$ verifies:

$$\lim_{t \to \infty} \varphi(y(t)) = \lim_{t \to \infty} \|y(t)\|^2$$

Moreover, if we call $\alpha(y(0))$ this limit, then the limit point of any convergent subsequence $y(t_k) \underset{t \to \infty}{\longrightarrow} \bar{y}$, is an eigenvector $\bar{y} \in P(\varphi, \varphi^*)$ with eigenvalue $\alpha(y(0)) \in \Lambda(\varphi, \varphi^*)$. ■

For norms induced by positive definite matrices Robert proved in [Ro] that his algorithm is reduced to the power method. Also see [Ga].

Example. The Power Method. [Ro] Let B be a symmetric and positive definite matrix. Consider $\varphi(y) = < y, By >^{1/2}$. Then $\gamma_\varphi(y) = \frac{By}{<y,By>^{1/2}}$. The dual norm is $\varphi^*(y) =< y, B^{-1}y >^{1/2}$, so the evolution (23) corresponds to:

$$y(t+1) = \frac{By(t)}{< y(t), By(t) >^{1/2}} \text{ with } y(0) \neq 0.$$

which is the power method. Any limit point $\bar{y} = \lim_{k\to\infty} y(t_k)$ satisfies $\alpha(y(0)) = < \bar{y}, B\bar{y} >^{1/2} =< \bar{y}, \bar{y} >$.

Since $\bar{y} = \lim_{k\to\infty} y(t_k+1)$ we get $\bar{y} = \frac{B\bar{y}}{<\bar{y},B\bar{y}>^{1/2}}$, so $<\bar{y},\bar{y}>= \frac{<B\bar{y},B\bar{y}>}{<\bar{y},B\bar{y}>}$. Then

$$< \bar{y}, B\bar{y} >=< \bar{y}, \bar{y} >< \bar{y}, \bar{y} >=< \bar{y}, \bar{y} > \frac{< B\bar{y}, B\bar{y} >}{< \bar{y}, B\bar{y} >}$$

We have shown the equality $< \bar{y}, B\bar{y} >= \|y\| \|B\bar{y}\|$, which implies $B\bar{y}$ and $\bar{y}$ are colinear. Then $\bar{y}$ is an eigenvector of B, $B\bar{y} = \alpha\bar{y}$ and its eigenvalue is $\alpha = \frac{\|B\bar{y}\|}{\|\bar{y}\|} = \frac{<\bar{y},B\bar{y}>}{\|\bar{y}\|^2} = \frac{(\varphi(\bar{y}))^2}{\|\bar{y}\|^2} = \varphi(\bar{y}) = \alpha(y(0))$. ■

We shall generalize Robert's algorithm. Let φ and ψ be continuous subnorms on the Hilbert space H and $\mathcal{H}$ respectively, and let φ^* and ψ^* be their extended dual subnorms. Let $B : H \to \mathcal{H}$ be a continuous operator and B^* be its adjoint operator. We assume that the following hypotheses hold:

$$\ker\varphi^* = B^{-1}(\ker\psi) \text{ and } \ker\psi^* = (B^*)^{-1}(\ker\varphi) \tag{4.13}$$

Let $x(0) \in H$ be such that $\varphi^*(x(0)) = 1$. By induction we define:

$$\ell(t) = \gamma_\psi(Bx(t)) \text{ and } x(t+1) = \gamma_\varphi(B^*\ell(t)) \text{ for } t \geq 0 \tag{4.14}$$

From Lemma 3.3 we deduce:

$$\psi^*(\ell(t)) = 1 \text{ and } \varphi^*(x(t)) = 1 \text{ for any } t \geq 0 \tag{4.15}$$

and,

$$\psi(Bx(t)), \leq \varphi(B^*\ell(t)) \leq \psi(Bx(t+1)) \quad \forall t \geq 0 \tag{4.16}$$

Then the sequences $(\psi(Bx(t)))$, $\varphi(B^*\ell(t))$ are increasing and their common limit is denoted by $\alpha(x(0))$:

$$\lim_{t\to\infty} \psi(Bx(t)) = \lim_{t\to\infty} \varphi(B^*\ell(t)) = \alpha(x(0)) \tag{4.17}$$

When φ and ψ are norms on $\mathbb{R}^n$ and $\mathbb{R}^m$ respectively, then they are continuous, as well as their dual norms φ^* and ψ^*. In this case $B : \mathbb{R}^n \to \mathbb{R}^m$ is induced by a matrix, and the dual operator is the transpose matrix B^t.

The sets

$$S_{\varphi^*} = \{x \in \mathbb{R}^n : \varphi^*(x) = 1\} \text{ and } S_{\psi^*} = \{\ell \in \mathbb{R}^m : \psi^*(\ell) = 1\}$$

are compacts. Then, from (4.16) and continuity of φ and ψ the quantity $\alpha(x(0))$ is finite.

Theorem 4.3. For any pair of norms, φ on $\mathbb{R}^n$ and ψ on $\mathbb{R}^m$, we consider the sequences defined by $\ell(t) = \gamma_\psi(Bx(t))$, $x(t+1) = \gamma_\varphi(B^t\ell(t))$, for $t \geq 0$, with $\varphi^*(x(0)) = 1$. Then $\lim_{t\to\infty} \psi(Bx(t)) = \lim_{t\to\infty} \varphi(B^t\ell(t))$. Moreover, if we call $\alpha(x(0))$ this limit, then there exists converging subsequences $x(t_r) \to \bar{x}$, $\ell(t_r) \to \bar{\ell}$ and the limit points verify:

$$\psi^*(\bar{\ell}) = 1 = \varphi^*(\bar{x}) \tag{4.18}$$

$$\psi(B\bar{x}) = \alpha(x(0)) = \varphi(B^t\bar{\ell}) \tag{4.19}$$

Proof. From compacity and (4.15) we deduce that there exists converging subsequences $(x(t_r))$, $(\ell(t_r))$. From continuity we get that their limits satisfy the set of equalities (4.18), (4.19). ■

5. Optimization Problem Associated with Neural Networks

Consider B a real $m \times n$ matrix, so its dual operator is B^t. From Theorem 3.3 we get:

$$\sup_{\ell \in \mathbb{R}^m : \psi^*(\ell)=1} \varphi(B^t\ell) = \sup_{x \in \mathbb{R}^n : \varphi^*(x)=1} \psi(Bx) \tag{5.1}$$

In the particular case $\psi(\ell) = \|\ell\|_2 = (\sum_{i=1}^{m} \ell_i^2)^{1/2}$ we get $\gamma_\varphi(\ell) = \frac{\ell}{\|\ell\|_2}$ and $\psi^* = \psi$. In this case the iteration (4.14) is reduced to the following one

$$\ell(t) = \frac{Bx(t)}{\|Bx(t)\|_2}, \; x(t+1) = \gamma_\varphi(B^t\ell(t)) \text{ for } t \geq 0, \tag{5.2}$$

with the initial point $x(0)$ verifying $\varphi^*(x(0)) = 1$. Hence $x(t)$ verifies the dynamical equation:

$$x(t+1) = \gamma_\varphi(\frac{B^tBx(t)}{\|Bx(t)\|_2})$$

From condition (2.3) we obtain that it is equivalent to:

$$x(t+1) = \gamma_\varphi(B^t Bx(t))$$

which is the evolution of a $\underline{P}$ (i.e. positive) neural network with symmetric interactions given by matrix $A = B^t B$. From (4.16) we get that $\|Bx(t)\|_2$ is an increasing functional for the evolution of this network.

Moreover from (4.16) we find that $(\varphi(x(t)) : t \geq 0)$ is an increasing functional for this evolution which corresponds to the functional that we introduce for this kind of networks in [GM].

Now, in this contex, take $\varphi(x) = \|x\|_1 = \sum_{i=1}^{n} |x_i|$. Then $\gamma_\varphi(x) =$ sign x and $\varphi^*(x) = \|x\|_\infty = \max\{|x_i| : i = 1, ..., n\}$. Hence (5.1) is reduced to:

$$\sup_{\ell \in \mathbb{R}^m : \|\ell\|_2 = 1} \|B^t \ell\|_1 = \sup_{x \in \mathbb{R}^n : \|x\|_\infty = 1} \|Bx\|_2 \tag{5.3}$$

and the iteration (5.2) corresponds to:

$$\ell(t) = \frac{Bx(t)}{\|Bx(t)\|_2}, \quad x(t+1) = \text{ sign } (B^t \ell(t)) \tag{5.4}$$

with initial point $x(0)$ satisfying $\|x(0)\|_\infty = 1$. Then $(x(t))$ verifies the evolution equation:

$$x(t+1) = \text{ sign } \left(\frac{B^t Bx(t)}{\|Bx(t)\|_2} \right) = \text{ sign } (B^t Bx(t)) \tag{5.5}$$

which is the evolution of a symmetric neural network with interation matrix $A = B^t B$.

From (5.5) we deduce that $x(t) \in \{-1, 0, 1\}^m$ for any $t \geq 1$. This immediatly implies that the sup in the problem:

$$(P) \qquad \sup_{x \in \mathbb{R}^n : \|x\|_\infty = 1} \|Bx\|_2$$

is attained in the set $\{-1, 0, 1\}^n$. In fact, suppose $\bar{x} \in \mathbb{R}^n$ satisfying $\|x\|_\infty = 1$ maximizes the problem (P). Take the initial point $x(0) = \bar{x}$ in (5.4). Since $\|Bx(t)\|_2$ increases with t, we deduce,

$$\|B\bar{x}\|_2 \leq \|Bx(1)\|_2$$

with $x_i(1) \in \{-1,0,1\}$. So, $x(1)$ also is an optimum of (P). Hence:

$$\sup_{x \in \mathbb{R}^n : \|x\|_\infty = 1} \|Bx\| = \sup_{x \in \{-1,0,1\}^n} \|Bx\| \tag{5.6}$$

By combining (5.3) and (5.6) we get the following result:

Theorem 5.1. [J] For any real $m \times n$ matrix B we have:

$$\sup_{x \in \{-1,0,1\}^n} \|Bx\|_2 = \sup_{\ell \in \mathbb{R}^m : \|\ell\|_2 = 1} \|B^t \ell\|_1 \tag{5.5}$$

■

Observe that this means that we can change a combinatorial quadratic problem into a continuous one. This last result was firstly obtained, by means of a different proof, in [J]. Recently Cominetti [C] has given another optimization theoretical frame which permits to obtain Theorem 5.1 as a corollary of powerful results.

6. Conclusions

We have analized automata networks as hill-climbing strategies for optimization problems. It follows from our study that the canonical optimization problem associated with automata networks corresponds to a norm optimization under a unit ball constraint. Moreover, the discrete quadratic problem which appears in the very beginning of automata approach to the spin glass optimization, can be seen in this framework; i.e. it is equivalent to a norm $\| \ \|_1$ optimization problem on the unit ball associated to the Euclidean norm.

Acknowledgments

We are indebted for fruitful discussion to A. Jofré and R. Cominetti. This work was partially supported by FONDECYT, under grants 1208-91 (SM), 1211-91 (EG), and by DTI, Universidad de Chile.

References

MR] Metropolis, N., A.W. Rosenbluth, M.N. Rosenbluth, A.M. Teller, E.J. Teller, J. Chem. Phys. 21, 1087 (1953).

[GJ] Garey, M.R., D.S. Johnson, *Computers and intractability.* Freeman, N. York (1979).

[K] Kirkpatrick, S., C.D. Gelatt, M.P. Vecchi, *Optimization by simulated annealing.* Science 220(4598), 671-680 (1983).

[LA] Van Laarhoven, P.J.M., E.H.L. Aarts, *Simulated annealing: theory and applications.* D. Reidel Publ., Kluwer (1987).

[MPV] Mezard, M., G. Parisi, M. Virasoro, *Spin glass theory and beyond.* World Scientific (1987).

[GM] Goles, E., S. Martínez, *Neural and automata networks.* Kluwer (1990).

[H] Hopfield, J.J., *Neural networks and physical systems with emergent collective computational abilities.* Proc. Natl. Acad. Sci. 79, USA, 2554-2558 (1982).

[HT] Hopfield, J.J., D.W. Tank, *Neural computation of decisions in optimization problems.* Biol. Cybernetic 52, 141-152 (1985).

[Ho] Householder, A.S., *The theory of matrices in numerical analysis.* Blaisdell Publ., New York, 1964.

[L] Laurent, J.P., *Approximation et optimisation*, Herrmann, Paris, 1972.

[S] Schwartz, L., *Analyse: topologie générale et analyse fonctionnelle*, Herrmann, Paris, 1970.

[Ro] Robert, F., *Calcul du rapport maximal de deux normes sur* $\mathbb{R}^n$, RIRO (5), 97-118, 1967.

[Ga] Gastinel, N., *Analyse numérique linéaire*, Herrmann, Paris, 1966.

[R] Rockafellar, R.T., *Convex Analysis*, Princeton Univ. Press, Princeton, N.J., 1970.

[Gi] Giles, J.R., *Convex analysis with application in differentiation of convex functions.* Pitman Adv. Publ. Prog., Boston, 1982.

[J] Janssen, A.J.E.M., *An optimization problem related to neural networks*, Information Processing Letters 37, 155-157, 1991.

[B] Barahona, F., *Application de l'optimisation combinatoire á certains modèles de verres de spin*, Doctoral Thesis, IMAG, Grenoble, France, 1980.

[PA] Peterson, C., J. Anderson, *Neural networks and NP-complete optimization problems; A performance study on the graph bisection problem*, Complex Systems, 2, 59-89, 1988.

[C] Cominetti, R., *Automata networks and optomization of differences of convex functions*, Preprint, 1990.

TWO CHOSEN EXAMPLES FOR FRACTALS: ONE DETERMINISTIC, THE OTHER RANDOM

H. J. HERRMANN
HLRZ, KFA Jülich
5170 Jülich
Germany

1. Space-Filling Bearings

1.1. INTRODUCTION

Let us ask if it is possible to tile a plane with wheels rolling on each other such that all the area is covered with wheels. This rather exotic question can arise in various contexts. One could imagine the wheels to be eddies on the surface of an incompressible fluid and then ask if the fluid motion can be totally decomposed into stable eddies. Or, one could think of mechanical roller bearings between two moving surfaces, like two tectonic plates, and then ask if one can completely fill the space between the rolling cylinders with other rolling cylinders such that no cylinder excerces any frictional work on another one. The question we are asking is, in fact, geometrical.

The original motivation for studying this problem was the enigmatic observation that over very extended areas, called "seismic gaps" [1], two tectonic plates can creep on each other without producing neither earthquakes nor the amount of heat expected from usual friction forces. One such region is a part of the San Andreas Fault that extends over more than one hundred kilometers between Los Angeles and San Francisco.

As a possible mechanism to explain this behaviour one could think that the material between the plates, which is called "gouge", organizes itself in such a way that it acts like a bearing. Since in a bearing one has no gliding friction but only rolling friction this could explain the lack of measurable heat production. On the other hand, in roller-skates for instance, mechanical bearings work because the individual balls or cylinders of the bearing are kept separate from each other by leaving rather big empty spaces between them. Under the pressure of kilobars that push tectonic plates against each other such empty spaces cannot exist. The space between the rolling cylinders must be filled with rolling matter such that the motion of the main cylinders is not hindered by gliding friction. So the main cylinders

E. Goles and S. Martínez (eds.), Statistical Physics, Automata Networks and Dynamical Systems, 89–118.

should roll on secondary cylinders which themselves roll on successive generations of smaller and smaller cylinders as shown in Fig. 1. Evidently provided such a bearing exists it can only be constructed iteratively and will therefore probably be self-similar†. Several seismic measurements have in fact indicated self-similar or turbulent motion within the gouge [2].

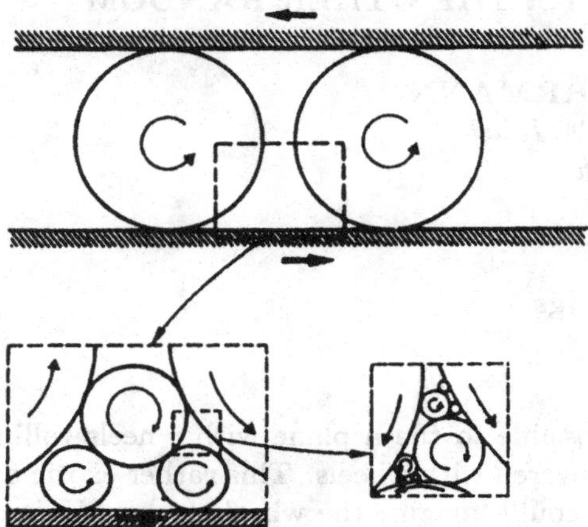

Figure 1. Schematic two-dimensional cut through a roller bearing between two tectonic plates. The inserts show how the holes could be filled with rotating cylinders.

In order for the above model to work in practice various conditions still need to be fulfilled: The individual stones within the gouge must be round; there must be a dynamics under which the system naturally evolves into the bearing; at some lower cut-off which can be given by the roughness of the surface of the stones another mechanism must take over and finally one has to justify considering cylinders or two-dimensional cuts instead of the full three-dimensional motion. It is not the aim of this course to deal with these questions. We want to concentrate just on the existence and construction of the space-filling self-similar bearings and study their geometrical properties.

Tiling space with circles by putting iteratively in each hole between three circles the circle that tangentially touches all three (see Fig. 2) is an old problem often known under the name of "Apollonian packing". It dates back to Apollonius of Perga who lived around 200 B.C. and much work has been done since as briefly presented for instance in Mandelbrot's book [3]. The space left over between circles is a fractal but despite much effort it has not yet been possible to determine the value of

† We will not discriminate in this course between self-similar and self-inverse [3].

the fractal dimension analytically. The best numerical estimate is $d_f \approx 1.3058$ [4].

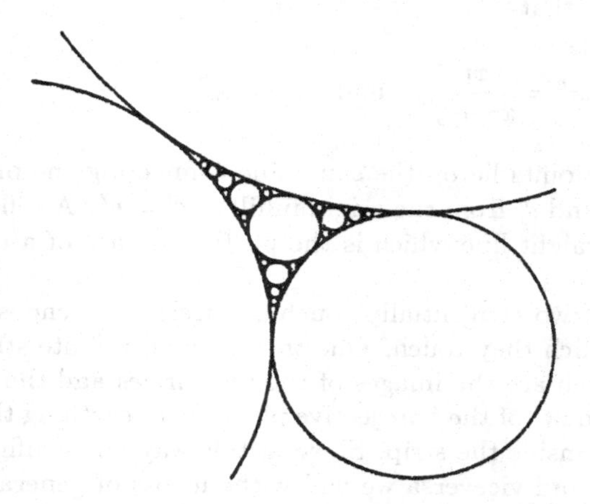

Figure 2. Apollonian packing.

The best way to construct Apollonian packings is by using iteratively circle-conser- ving mappings (Möbius transformations) [5]. This technique turns out to be also suited to our problem of bearings. We will in the following section describe how these mappings work. Using their properties we derive in the following section a necessary condition for the existence of a solution with fourfold loops and show via explicit construction that this condition is also sufficient. The section that follows concentrates on fractal dimensions and size distributions of the packings. Finally we conclude by comparing our results to possible physical applications. The work presented in this course will be published in Ref. 6.

1.2. THE CONSTRUCTION OF SELF-SIMILAR PACKINGS

Möbius transformations are conformal (i.e. angle-conserving) two-dimensional mappings that map circles into circles. In the complex z-plane they are in general given by

$$z' = \frac{a + bz}{c + dz} \tag{1}$$

where the constants a, b, c and d fulfill $ad - bc = 1$. Möbius transformations can be decomposed into translations, rotations, reflexions and inversions. Only the inversions, however, change the size of the circles and will therefore constitute the central element in the iteration of contracting mappings that we need.

The inversion around a circle of radius r centered at the origin maps point (x, y) into (x', y') such that

$$x' = \frac{xr^2}{x^2 + y^2} \quad \text{and} \quad y' = \frac{yr^2}{x^2 + y^2} \quad . \tag{2}$$

In other words, both points lie on the same line connecting the origin to infinity and their distances d and d' from the origin fulfill $d \cdot d' = r^2$. A reflexion is just an inversion around a straight line which is the particular case of a circle of infinite radius.

Suppose one has two tangentially touching circles and chooses as inversion center the point at which they touch. One then gets an infinite strip bounded by two paralle l lines which are the images of the two circles and the touching point becomes the infinite point (of the "projective plane"). Everything that was outside the two circles is now inside the strip. Since in this way any configuration can be mapped on to a strip and viceversa we will without loss of generality consider in the following only strip geometries.

Strips are very well suited for our purpose of constructing packings because they have the following property: Suppose one has a network of mutually touching circles within the strip. Then each inversion around a point at which two circles tangentially touch will map the strip into another strip preserving the topology of the network. In particular, if the original network was self-similar the new strip will be identical to the original one except for an eventual change in width. In this way the strip geometry is invariant under the most interesting kind of inversions.

Evidently the main problem one has to solve in order to construct space-filling packings is that of filling the wedge between two tangentially touching circles. This can be achieved on a strip by an inversion around the touching point one wants to fill because in this way the entire strip is mapped into the wedge. The strip itself is much easier to construct because one can systematically fill it by periodically repeating a given network of circles. This can be understood best by explicitely illustrating the construction mechanism in Fig. 3.

The circle A (the biggest one) is placed inside the strip and an inversion is made around the dashed circle (Fig. 3a) giving A'. At each iteration step the ensemble of circles is shifted to the right by a and a new circle A (or B) must be placed each time (Figs. 3a and 3b). Instead of shifting the ensemble of circles to the right one can of course also shift the inversion circle to the left which is actually what is shown in Fig. 3. All the other circles are the product either from an inversion around a dashed circle or a reflexion around a dashed line. The positions of the center of inversion and of the placed circles alternate between top and bottom. In fact we have called the placed circles A when they are on top and B when they are on the bottom of the strip. The essential condition for space-filling seems to be that the light gray area in Fig. 3d be mapped in such a way that its image-areas (dark gray areas) precisely fill the space created by the shift by a and not covered by circle A without that these image-areas overlap. In order to fulfill these conditions one has to carefully choose the radius R of A, the period $2a$ and the radius r_A of the inversion circle. It will be the subject of the next section to analytically derive expressions for their values.

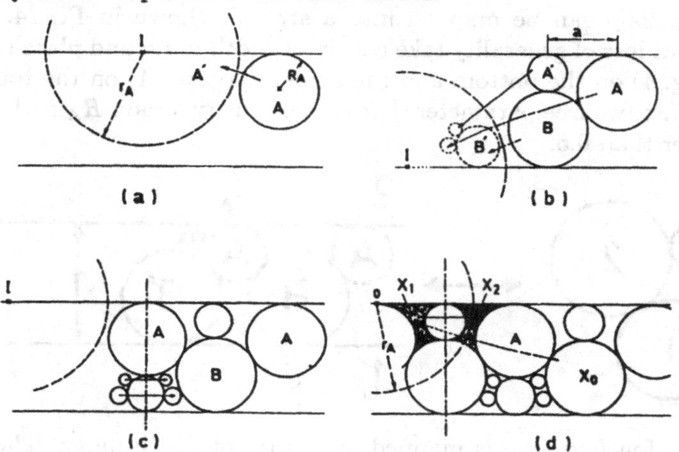

Figure 3. Schematic representation of how to construct a space-filling self-similar packing on a strip. In (d): $X_2 = \mathcal{R}X_1 = \mathcal{R}\mathcal{I}X_0$.

The other important condition we must fulfill is the slipless rotation of each disc on its tangential neighbors. Discs can rotate either clockwise or counterclockwise, i.e. there are two types of disks. A clockwise rotating disc can only touch counterclockwise rotating ones and viceversa. Consequently any loop of touching discs one can form in the packing must have an even number of discs. Suppose one constructs one of these loops by starting with one disc and adding one by one the discs of the loop. If the first disc rotates with a tangential velocity v its touching neighbor will have the same tangential velocity due to the slipless motion at the contact point. So all the disks in the loop will have the same tangential velocity and therefore when one closes the loop with the last disc one will not encounter any slip at the two contact points. We have shown that in fact any loop of even number of discs will fulfill the requirement that all discs can rotate siplessly on each other having all the same tangential velocity. In the same spirit one can now attach other loops to the existing loop and convince oneself that if the packing has only loops of even number of disks it suffices to turn any one of the discs and all the discs will start to rotate sliplessly on each other with the same tangential velocity v. The problem is therefore reduced to the construction of packings with only even loops.

1.3. THE COMPLETE SET OF SOLUTIONS WITH FOURFOLD LOOPS

Let us consider in the following only loops of length four. As we discussed above any such loop can be mapped into a strip as shown in Fig. 4. We can therefore without loss of generality take a strip of width unity and place the largest circle (B in Fig. 4) on the bottom and the second largest (A) on the top. We see that there are just two free parameters left namely the two radii R_A and R_B which cannot be larger than 0.5.

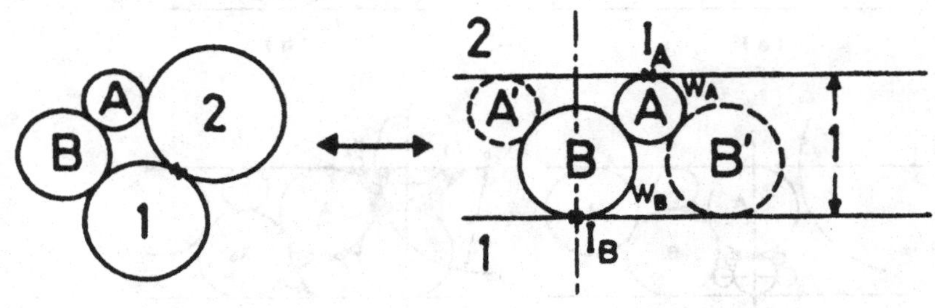

Figure 4. Any fourfold loop is mapped on a strip of width unity. There wedge w_B (w_A) is filled by an inversion around I_B (I_A). Consequently one has a reflexion symmetry around the dashed line and the dotted circle A' must exist.

As described in the previous section we have to fill in the wedges by inversions and since our mapping is contracting we must start by filling the largest wedge which is the one between B and the border of the strip (w_B in Fig. 4). The center of inversion is the touching point I_B. Since straight lines leaving I_B are mapped into straight lines and since the border of the strip must remain smooth after the inversion we necessarily have a reflexion symmetry around the dashed line in Fig. 4.

Due to this reflexion symmetry an image of A must exist to the left which in Fig. 4 is the dotted A' if the circle A is not invariant under reflexion. In this case, which we call "first family", one can by applying the same argument also to wedge w_A at circle A establish that a circle B' must touch A on the right. So one finds a strip of mutually touching largest circles with a period of $2a$ as shown in Fig. 5a.

On the other hand, there exists also the "second family" for which A and B are both invariant under the same reflexion. In this case the largest circles will not necessarily touch each other. Due to self-similarity one expects, however, still to find eventually a finite period $2a$ (see Fig. 5b); we come to talk about irregular solutions at the end of this section. In this case there exists a straight line tangentially connecting all the touching points of the largest circles A and B (dashed line in Fig. 5b). This line is mapped by both inversions around I_A and I_B on the same circle. Therefore a point C exists on the dashed line through which both inversion circles go.

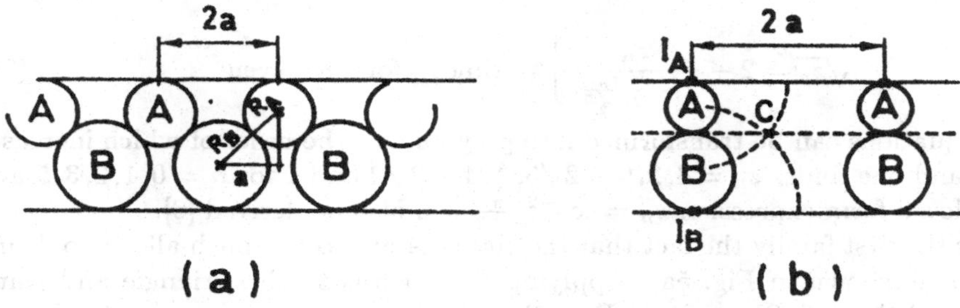

Figure 5. Schematic position of the largest circles in (a) the first family, (b) the second family.

Since in an inversion wedges must be mapped into wedges and using that our strip has width unity one finds the general relation

$$2R_A = r_A^2 \quad \text{and} \quad 2R_B = r_B^2 \tag{3}$$

between the radii $R_{A,B}$ of circles A, B and the radii $r_{A,B}$ of the inversion circles centered at $I_{A,B}$.

Let us now focus on the necessary condition for the mapping to be space-filling. We have seen from Fig. 3d that the areas are filled via sequences of inversions $\mathcal{I}$ and reflexions $\mathcal{R}$. If one just regards now the points at which the images of the circle A touch the border of the strip one sees that their distances x from the inversion center (which for simplicity we put on the origin) are mapped like[†]:

$$\mathcal{I} : x \to \frac{r^2}{x} \quad \text{and} \quad \mathcal{R} : x \to 2a - x \quad . \tag{4}$$

Since the touching point x_0 of circle A with the border is $2a$ the n-th iterate x_n is given by

$$x_n = \mathcal{IRIRI}(2a) \quad . \tag{5}$$

Either the iteration is infinite or it terminates after n steps: For even n the last circle is invariant under inversion and $x_n = r_A$, the radius of the inversion circle. For odd n the n-th circle is invariant under reflexion: $x_n = a$. Using Eq. (4) we have

$$\mathcal{IR}(x) = \frac{r_A^2}{2a - x} = \frac{az}{2 - \frac{x}{a}} \quad \text{with} \quad z = \frac{r_A^2}{a^2} \tag{6}$$

[†] Let us note that $\mathcal{RI}$ is holomorphic although $\mathcal{I}$ and $\mathcal{R}$ are not.

which inserted in Eq. (5) gives for the fixed point of the iteration the equations for z

$$1 = \left.\frac{z_n}{2-\frac{z_n}{2-\frac{z_n}{2\ldots}}}\right\}\frac{n+1}{2} \quad \text{times} \quad \text{for} \quad n \quad \text{odd} \tag{7a}$$

and

$$\sqrt{z_n} = 2 - \left.\frac{z_n}{2-\frac{z_n}{2-\frac{z_n}{2\ldots}}}\right\}\frac{n}{2} \quad \text{times} \quad \text{for} \quad n \quad \text{even} \quad . \tag{7b}$$

These equations can be transformed into polynomials the order of which increases with n and one finds: $z_n = 4, 2, 6-2\sqrt{5}, \frac{4}{3}, 4-2\sqrt{2}$ and 1 for $n = 0, 1, 2, 3, 5$ and ∞. A closed form expression $z_n = \cos^{-2}\frac{\pi}{n+3}$ can also be derived [6].

For the first family the fact that the circles A and B all touch allows to define the triangle shown in Fig. 5a. Applying Pythagoras to this triangle and using Eq. (3) and the definition of z in Eq. (6) we find

$$a^{-2} = z_n + z_m - 1 \quad \text{and} \quad r^2_{A,B} = \frac{z_{n,m}}{z_n + z_m - 1} \tag{8}$$

where n (m) is the number iterations made around I_A (I_B), i.e. the number of images one makes of A (B); n and m are integers 0,1,...,∞. For each pair (n, m) there exists at best one solution and for this solution the period $2a$, the radii of inversion $r_{A,B}$ and the radii of the largest circles $R_{A,B}$ are fixed, specific numbers. Similarly, for the second family, one exploits the fact that in Fig. 5b point C lies on the dashed line and on both inversion circles and one obtains

$$a^{-2} = z_n + z_m \quad \text{and} \quad r^2_{A,B} = \frac{z_{n,m}}{z_n + z_m} \quad . \tag{9}$$

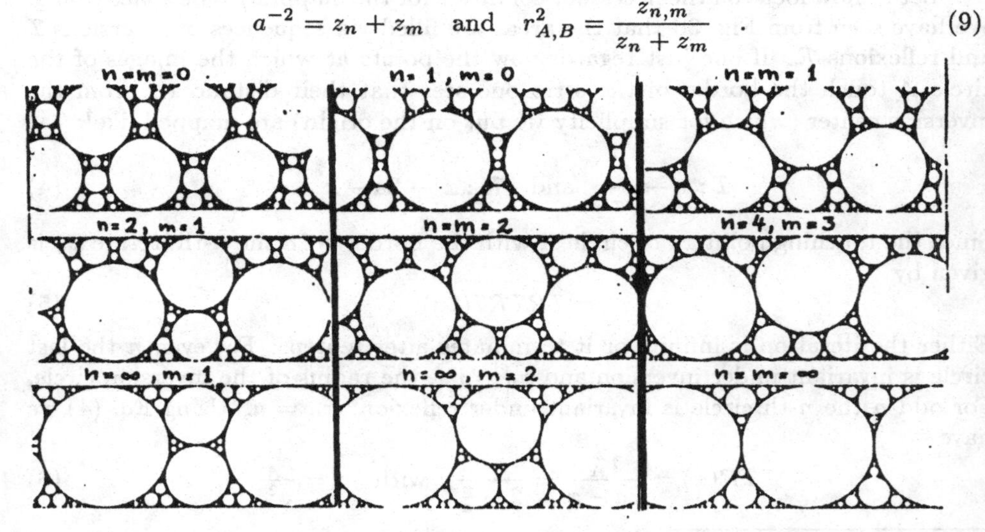

Figure 6. Nine different combinations of n and m of the first family.

In order to see if the packings of Eqs. (8) and (9) really exist we construct them explicitely. For the first family we use the algorithm described in the previous section; in Fig. 6 we see some examples. It is interesting to note that for $n = m = \infty$ and for $n = 0, m = \infty$ one obtains the classical Apollonian packing. In fact the whole one-parameter family with $n = \infty$ and arbitrary m has threefold loops and is in this sense the full "Apollonian family". It is obtained in this case as the limit in which one of the four circles in each loop becomes infinitely small.

The explict construction of the second family is more complicated and a simple algorithm is not yet known [6]. In Fig. 7a we see the case $n = m = 0$. If $nm + n + m > 3$ $(n, m \neq \infty)$ the solutions of the second family have circular holes, the largest of which has radius $\hat{R} = \sqrt{z_n + z_m - z_n z_m/(z_n + z_m)}$, which only touch infinitesimally small circles (powder). The case $n = \infty$ of the second family is identical to the case $n = 1$ of the first family for any fixed m.

We have now found all the regular solutions with fourfold loops. It is, however, also possible to make mixtures between them by putting one into the other. Since the only shape two solutions have in common is the circle the mixing is best done by conformally mapping the strip into a circular geometry. In Fig. 7b we see the case $n = m = 4$ of the first family mapped into such a geometry (by inverting the strip around its circle B). Each solution (n, m) can be mapped in this way into a circular geometry and then one can replace the circles of a given regular packing by randomly chosen packings in circular geometry. This is the most general packing of fourfold loops one can construct.

Figure 7. Bearings of fourfold loops (a) $n = m = 0$ of the second family in strip geometry; (b) $n = m = 4$ of the first family in circular geometry.

1.4. FRACTAL DIMENSIONS

The packings constructed in the last section are evidently fractal. One way to define their fractal dimension is by introducing a cut-off length ϵ such that one considers in a packing exactly those circles that have a radius larger than ϵ. Now one can calculate (on the computer) the number $N(\epsilon)$ of circles per unit area, the sum $s(\epsilon)$ of the perimeters of the circles ("surface") per unit area and the "porosity" $p(\epsilon)$, i.e. the area that is not covered by circles per unit area. All these quantities can be related to the distribution $n(r)$ of radii r, i.e. the number of circles of radius r per unit area through:

$$N(\epsilon) = \int_{\epsilon}^{\infty} n(r)dr \tag{10a}$$

$$s(\epsilon) = 2\pi \int_{\epsilon}^{\infty} rn(r)dr \tag{10b}$$

$$p(\epsilon) = 1 - \pi \int_{\epsilon}^{\infty} r^2 n(r)dr \quad . \tag{10c}$$

If $n(r)$ can be described by a simple power law

$$n(r) \sim r^{-\tilde{\tau}} \tag{11}$$

then one finds

$$N(\epsilon) \sim \epsilon^{-d_f}, \quad s(\epsilon) \sim \epsilon^{1-d_f} \quad \text{and} \quad p(\epsilon) \sim \epsilon^{2-d_f} \quad \text{with} \quad d_f = \tilde{\tau} - 1 \tag{12}$$

where d_f is the fractal dimension [3].

In Fig. 8 we show N, s and p plotted double logarithmically against the cut-off for two cases. The straight lines over several orders of magnitude confirm the power-law behaviour. The fact that the porosity goes to zero with ϵ is a numerical verification that the packings are space-filling. The fractal dimension one obtains from the porosity for the Apollonian packing, i.e. $n = m = \infty$ agrees well with Boyd's value [4] and seems to be the same for the whole Apollonian family. For the first family the fractal dimension obtained from the porosity continuously increases with decreasing n and m and is 1.42 for $n = m = 0$. For the second family one finds 1.52 for $n = m = 0$.

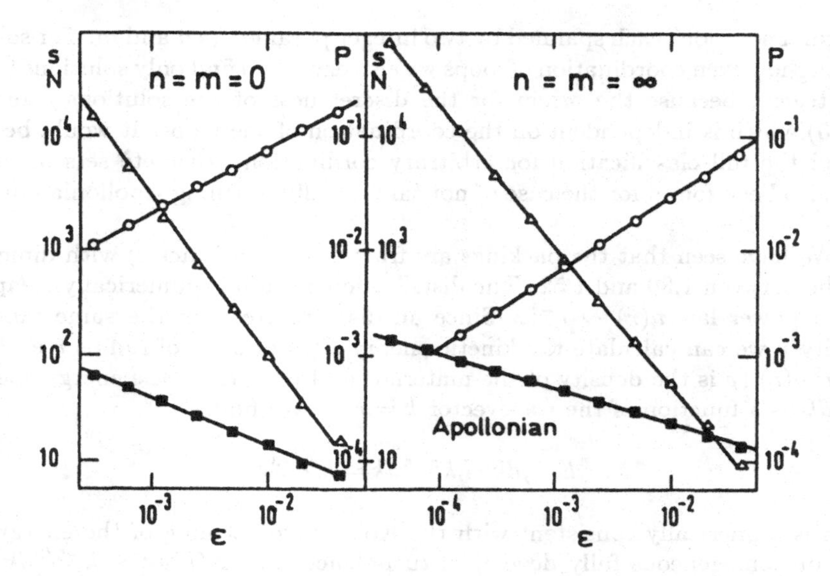

Figure 8. Log-log plot of the number of circles N, the surface s and the porosity p as function of the cut-off ϵ for the first family. The fractal dimensions d_f obtained from N, s and p are: 1.47, 1.45 and 1.42 for (a) and 1.33, 1.33 and 1.30 for (b).

¿From Fig. 8 one also makes the somewhat startling observation that numerically Eq. (12) seems to be violated because the fractal dimensions one obtains from the various moments of Eq. (10) do slowly decrease with increasing order of the moment. This is due to numerical inaccuracies: Through large scale computations and taking into account curvatures coming from corrections to scaling it has been found that they are asymptotically the same [7]. It could, however, also be possible that different moments scale differently due to multifractality [8] which has in fact been observed when instead of looking at all circles larger than ϵ one takes all circles produced up to a given level of the iteration multifractality of the Apollonian packing [9]. Multifractality has was also found by looking of various moments of the density of contact points between circles taking all circles up to a given iteration [10]. at all circles It would be interesting to investigate this point further.

1.5. SUMMARY AND DISCUSSION

We have found that space-filling packings with only fourfold loops in which each disc rotates sliplessly on its neighbors do exist and that there are two families of regular solutions each spanned by two integer parameters n and m. For solutions with higher, even coordination of loops we also expect to find only solutions labelled by integers, because the origin for the discreteness of the solutions came from Eq. (5) which is independent on the coordination of the loops. It would be useful to find the full classification for arbitrary cordination. Discrete sets of solution have also been found for the case of not tangentially touching Apollonian packings [11].

We have seen that the packings are fractal (or multifractal) with dimensions that lie between 1.30 and 1.52. The distribution of radii is numerically compatible with a power law $n(r) \sim r^{-\bar{\tau}}$. Since all discs rotate with the same tangential velocity v we can calculate the kinetic energy $E(r)$ of discs of radius r as $E(r) = \pi \rho v^2 r^2 n(r)$ (ρ is the density of the material) and from there the energy spectrum $E(k)dk$ as a function of the wavevector $k = r^{-1}$. We find:

$$E(k)dk \sim k^{\bar{\tau}-4} dk = k^{d_f - 3} dk$$

which is numerically consistent with the Kolmogoroff scaling of the energy spectrum of homogeneous fully developed turbulence [12]: $E(k)dk \sim k^{-5/3}dk$. This interesting coincidence might have a more profound meaning and might give a geometrical interpretation of turbulence in the picture of the transfer of energy to smaller and smaller eddies in the inertial regime.

In turbulence, however, the eddies interpenetrate and the fluid does not move like a solid block. Therefore a more direct realization of our rolling packings can be given by an experiment in which one rotates a fluid between two (excentric) cylinders like shown in Fig. 7b. If the position and size of the inner cylinder corresponds to one of our solutions one could expect the fluid to convect along the smaller intermediate circles in Fig. 7b. Of course, experimentally one still has to compensate for the viscosity of the fluid. Let us point out that, Fig. 7b could also be seen as a cross section of a mechanical excentric roller bearing.

For our original application to tectonic plates, however, still a problem remains: In all our solutions with fourfold loops on a strip the two borders move in the same direction (see e.g. Fig. 7a) instead of moving in opposite directions as tectonic plates would do. It is actually likely that even for higher coordinations or a mixing of them it will not be possible to find solutions with boundaries moving in opposite sense. However, if the boundaries are not perfectly straight but have some finite curvature one can produce solutions using conformal mappings like seen in Fig. 7b. Further investigation in this direction still needs to be done [6].

2. Damage Spreading

2.1 CELLULAR AUTOMATA

Cellular automata are dynamical systems with many discrete degrees of freedom. Each cell that contains one discrete variable has a rule which defines its state at the next time step as a function of the values of the neighboring cells and the cell itself. Cellular automata produce patterns in space and time that can be moving fractal clusters if the dynamics is chaotic.

To see if the dynamics of a cellular automaton is chaotic one watches how easily an error or "damage", i.e. a small change in the configuration spreads. In models with a tunable parameter one can find a transition between a frozen phase, in which the damage heals and a chaotic phase, in which the damage spreads to infinity. One example is the Kauffman model [13,14] where the transition can be located using the gradient method [15] by letting a damage front spread in a gradient. Another example are thermal Monte Carlo simulations which will be discussed extensively here.

The rule of a cellular automaton can be deterministic or probabilistic. The first case has been studied intensively [16,17] for binary states, e.g. 0 and 1. Although deterministic they can be disordered, either in the choice of the rule, like in the Kauffman model [13,14], or in the initial configuration like in Q2R [18]. In all cases binary, deterministic automata can be coded in the computer as one-site one-bit ("multi-spin-coding") and then as many sites treated in parallel as there are bits in a computer word using logical bit-by-bit operations. In addition the inner loops of the programs can be vectorized easily so that very high performances in speed can be achieved (1.1 GHz and 3.2 GHz for Q2R [19] on a Cray 2 processor and a full CM-2 respectively and 1.7 GHz for Q2R and 200 MHz for the Kauffman model [20] on one processor of a a Cray-YMP). Also very big systems can be simulated due to the multi-spin-coding (1.5×10^{10} sites for Q2R [19]).

Probabilistic automata use random numbers at each time-step. Examples are the Hamiltonian formulation of the Kauffman model [21] and annealed models [14,22]. But the most prominent example is thermal Monte Carlo dynamics to which the rest of this course is devoted.

2.2. DAMAGE SPREADING IN MONTE CARLO

Monte Carlo can be viewed as a dynamical process in phase space, i.e. starting from a given initial configuration one follows on a trajectory in phase space under the application of the Monte Carlo procedure. This trajectory will of course depend on the specific type of Monte Carlo (heat bath, Glauber, Metropolis, Kawasaki, etc) but also on the sequence of random numbers. The trajectories will always go towards equilibrium which means that the configurations will be visited by the trajectory with a probability proportional to the Boltzmann factors. Once in equilibrium the trajectories will stay there as assured by the detailed-balance condition.

We want to ask the question if such a dynamics is chaotic. More precisely: Suppose we make a small perturbation in the initial condition will the new trajectory be just slightly different or will it be totally different. The second case in which two initially close trajectories will become very fast very different is generically called chaotic. The detailed definition of chaos may in some cases also include the speed with which the trajectories separate but in our context we do not want to make this distinction.

In order that the concept of closeness of trajectories be meaningful one must define what is a distance in phase space. Suppose we consider a system of Ising variables then a useful metric can be given by the "Hamming distance" or "damage"

$$\Delta(t) = \frac{1}{2N} \sum_i |\sigma_i(t) - \rho_i(t)| \tag{13}$$

where $\{\sigma_i(t)\}$ and $\{\rho_i(t)\}$ are the two (time-dependent) configurations in phase space and N is the number of sites, labelled by i. This definition is certainly arbitrary and the results that we present in the following depend on this definition. Physically it just measures the fraction of sites for which the two configurations are different. The definition of Eq. (13) can easily be generalized to continuous variables [23] or to variables of q-states [24].

According to the discussion above one would then call a dynamical behaviour chaotic when in the thermodynamic limit $\Delta(t)$ goes to a finite value for large times if $\Delta(0) \longrightarrow 0$. The opposite of chaotic is called frozen, i.e. when $\Delta(\infty) = 0$ if $\Delta(0) \longrightarrow 0$. If one wants to get the limit $\Delta(0) \longrightarrow 0$ properly in a numerical simulation one can do the following [25]: Consider three configurations $\{\sigma_A\}$, $\{\sigma_B\}$ and $\{\sigma_C\}$ with $\Delta_{AB}(0) = \Delta_{BC}(0) = \frac{1}{2}\Delta_{AC}(0) = s$ where s is a fixed small number and Δ_{AB} denotes the distance between $\{\sigma_A\}$ and $\{\sigma_B\}$. Then

$$\Delta(t) = \Delta_{AB}(t) + \Delta_{BA}(t) - \Delta_{AC}(t) \tag{14}$$

is a very good extrapolation to $\Delta(0) \longrightarrow 0$.

The crucial idea in order to study damage spreading in Monte Carlo is therefore to apply on two configurations the *same sequence of random numbers* in the Monte Carlo algorithm. In this way one is taking the same dynamics for both configurations. To get statistically meaningful results one must then average over many initial configurations of equal initial distance and over many different sequences of random numbers. Also note that, as in deterministic dynamical systems, once the two configurations become identical they will always stay identical.

2.3. NUMERICAL RESULTS FOR THE ISING MODEL

In order to see if Monte Carlo can generate chaotic behaviour in the sense outlined above Glauber dynamics was applied in Ref.25 on the two dimensional Ising model. Using Eq.(14) the distance of initially close equilibrium configurations was calculated after 10^4 updates per site, i.e. after a long time. The result is shown in Fig. 9 as a function of temperature. Let us remark that since the calculation was performed in a finite system of size L the final distance would have vanished after a time of the order of $\exp\left(L^2\right)$ because eventually two uncorrelated trajectories in a finite phase space will always meet. The times considered in Fig. 9 are, however, much smaller than these "Poincaré" times. We see from Fig. 9 that above a certain temperature T_s the Glauber dynamics is chaotic while below T_s it is frozen. The order parameter for this transition between chaotic and frozen is the quantity plotted in Fig. 9. The data indicate that T_s is very close if not identical to the critical temperature T_c of the Ising model. It has not been possible up to now to determine the critical exponent β of this transition because the statistical fluctuations are very strong.

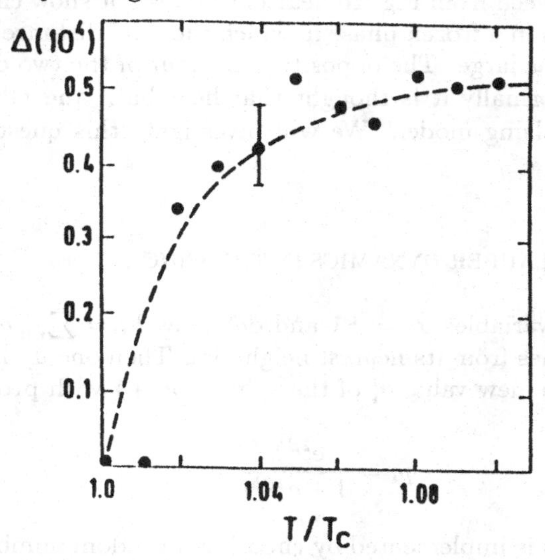

Figure 9. Hamming distance 10^4 time step for $\Delta(0) \longrightarrow 0$ using Glauber dynamics as a function of T/T_c for the two dimensional Ising model (from Ref. 25).

In three dimensions using Glauber dynamics qualitatively the same picture emerges as in two dimensions only the temperature T_s seems to be 4% below T_c as found in Ref.26. Particularly interesting is the fact that one finds a transition line even in a homogeneous field [27]. This indicates clearly that the dynamic transition between chaotic and frozen is not identical to the transition between the ferromagnetic and the paramagnetic phase. The transition line neither agrees with the percolation transition of minority spins [27]. It is thus an open question whether this dynamical transition is related to any known property of the Ising model or if it is a novel phenomenon.

The same questions have been addressed using a slightly different dynamics, namely heat bath [28]. The fraction $P(t)$ of the pairs of configurations that had not yet become identical after a time t and the average distance $D(t)$ between only those not yet identical configurations were calculated, $(\Delta(t) = P(t) \cdot D(t))$. In Fig. 10 we show the result with not thermalized configurations and looking at different values of initial damage. After a time of 500 updates per site. One sees that the survival of damage depends on the initial damage (Fig. 10a) while when the configurations are different their distance does not depend on the initial distance (Fig. 10b). We also see that if the initial distance goes to zero the final distance is also going to vanish because of $P(t)$. This is in agreement with the result found for Glauber dynamics in the ferromagnetic phase. In the paramagnetic phase, however, the result for heat bath is strikingly different from the Glauber dynamics because as seen from Fig. 10 heat bath does not show chaotic behaviour but on the contrary is in a frozen phase in which the final distance vanishes even if the initial distance was large. The opposite behaviour of the two dynamics is very surprising because normally it is thought that heat bath and Glauber dynamics are identical for the Ising model. We will investigate this question in the next section.

2.4. HEAT BATH VS GLAUBER DYNAMICS IN THE ISING MODEL

Let us consider variables $\sigma_i = \pm 1$ and define as $h_i = \sum_{nn} \sigma_j$ the local field acting on σ_i that comes from its nearest neighbors. Then one update in heat bath is given by setting the new value σ_i' of the spin to be $+1$ with probability p :

$$p_i = \frac{e^{2\beta h_i}}{1 + e^{2\beta h_i}} \quad . \tag{15a}$$

On the computer this is implemented by choosing a random number $z \in [0, 1]$ and setting

$$\sigma_i' = \text{sign}\,(p_i - z) \quad . \tag{15b}$$

In Glauber dynamics a spin is flipped with a probability

$$p(\text{flip}) = \frac{e^{-\beta \Delta E}}{1 + e^{-\beta \Delta E}} \tag{16a}$$

where ΔE is the difference between the energy of the would-be new configuration and of the old configuration. For the Ising model one has $\Delta E = 2\sigma_i h_i$. On the computer one implements Glauber dynamics via

$$\sigma_i' = -\sigma_i \ \text{sign}(p(\text{flip}) - z) \tag{16b}$$

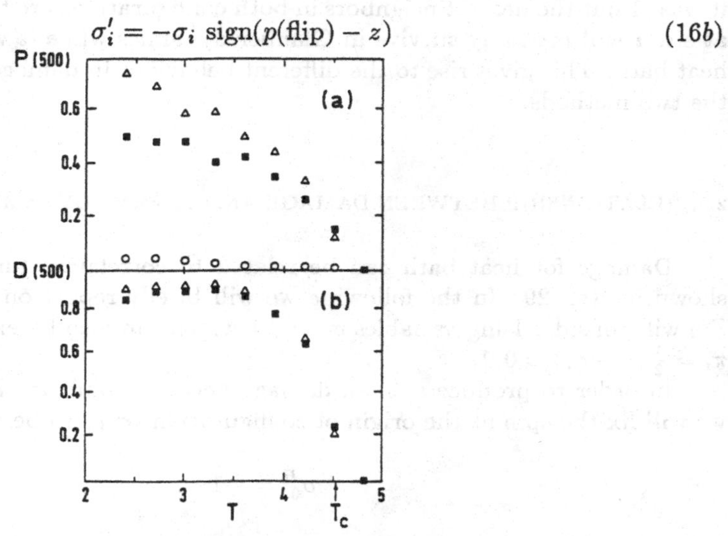

Figure 10. Damage after 500 time steps in the 3d Ising model as a function of temperature for $L = 12$ from Ref.28 for various initial damages: $\Delta(0) \longrightarrow 0$ (o), $\Delta(0) = \frac{1}{2}$(□) and $\Delta(0) = 1(\Delta)$; (a) fraction $P(t)$ of non-identical configurations and (b) distance $D(t)$ between two configurations provided they are still different.

Using Eqs. (15a) and (16a) p(flip) can be expressed in terms of p_i by

$$p(\text{flip}) = \begin{cases} 1 - p_i & \text{if} \quad \sigma_i = +1 \\ p_i & \text{if} \quad \sigma_i = -1 \end{cases}$$

and (17)

$$p(\text{not flip}) = \begin{cases} p_i & \text{if} \quad \sigma_i = +1 \\ 1 - p_i & \text{if} \quad \sigma_i = -1 \end{cases}$$

One sees from Eq. (17) that for the Glauber dynamics σ_i' is set +1 with probability p_i and -1 with probability $1 - p_i$ just as in heat bath so that both dynamics have exactly the same probabilities.

Inserting Eq. (17) into Eq. (16b) one finds how Glauber dynamics is implemented on the computer:

$$\begin{aligned} \sigma_i' &= \text{sign}\,(z-(1-p_i)) \quad \text{if} \;\; \sigma_i = +1 \\ \sigma_i' &= \text{sign}\,(p_i - z) \qquad\quad\; \text{if} \;\; \sigma_i = -1 \end{aligned} \tag{18}$$

This means that depending on the value of σ_i one uses the random number differently. So, if in one configuration the site σ_i was +1 and in the other configuration it was -1 but the nearest neighbors in both configurations are the same the damage at site i will probably survive in Glauber dynamics while it will certainly heal in heat bath. This gives rise to the different behaviour in damage spreading between the two methods.

2.5. RELATIONSHIP BETWEEN DAMAGE AND THERMODYNAMIC PROPERTIES

Damage for heat bath can be related to correlation functions as has been shown in Ref. 29. In the following we will briefly report on these relationships. We will consider Ising variables $\sigma_i = \pm 1$ which can also be expressed as usual by $\pi_i = \frac{1}{2}(1-\sigma_i) = 0, 1$.

In order to produce a small damage between configurations $\{\sigma_A\}$ and $\{\sigma_B\}$ we will fix the spin at the origin of configuration $\{\sigma_B\}$ to be always

$$\sigma_0^B = -1 \quad . \tag{19}$$

This represents thus a source of damage. In principle two types of damage can be imagined at site i : damage $+-$ where $\sigma_i^A = +1$ and $\sigma_i^B = -1$ or damage $-+$ where $\sigma_i^A = -1$ and $\sigma_i^B = +1$. The probabilities of finding a certain type of damage at site i can then be expressed as:

$$d_i^{+-} = \langle\langle (1-\pi_i^A)\,\pi_i^B \rangle\rangle \;\; \text{and} \;\; d_i^{-+} = \langle\langle \pi_i^A\,(1-\pi_i^B) \rangle\rangle \tag{20}$$

where $\langle\langle \cdots \rangle\rangle$ denotes a time average. Let us define the difference between the damage and use Eq. (20):

$$\Gamma_i = d_i^{+-} - d_i^{-+} = \langle\langle \pi_i^A \rangle\rangle - \langle\langle \pi_i^B \rangle\rangle \tag{21}$$

where we have used Eq.(10).

We want to express the damage through thermodynamic quantities defined on an unconstrainted system $\{\pi_i\}$ with averages taken over many configurations.

So we translate condition (19) by a conditional probability and use ergodicity to go from time averages to thermal averages:

$$\langle\langle \pi_i^A \rangle\rangle = \langle \pi_i \rangle \quad \text{and} \quad \langle\langle \pi_i^B \rangle\rangle = \frac{\langle \pi_i (1 - \pi_0) \rangle}{\langle 1 - \pi_0 \rangle} \tag{22}$$

where $\langle \cdots \rangle$ denotes a thermal average. Inserting this into Eq. (21) one finds

$$\Gamma_i = \frac{\langle \pi_i \pi_0 \rangle - \langle \pi_i \rangle \langle \pi_0 \rangle}{1 - \langle \pi_0 \rangle} = \frac{C_{0i}}{2(1-m)} \tag{23}$$

where

$$C_{0i} = \langle \sigma_i \sigma_0 \rangle - \langle \sigma_0 \rangle^2 \quad \text{and} \quad m = \langle \sigma_0 \rangle \tag{24}$$

are just the correlation function and the magnetization. Relation (23) gives us an equality of a certain combination of the damages with thermodynamical functions. In its derivation we used ergodicity but did not make any assumptions on the dynamics, the random numbers or the type of interaction and it is therefore of a very general validity.

If one would have chosen another form of fixed damage the result would have changed slightly. For instance fixing $\sigma_0^B = -1$ and $\sigma_0^A = +1$ one would find $\Gamma_i = C_{0i} / (1 - m^2)$.

If one considers variables with more degrees of freedom than Ising variables things can become more complicated but are still in principle feasible. As an example let us look at the Ashkin-Teller model [30] where on each site one has two binary variables $\sigma_i = \pm 1$ and $\tau_i = \pm 1$ and a Hamiltonian per site

$$\mathcal{H}_{ij} = -K(\sigma_i \sigma_j + \tau_i \tau_j) - 2L\, \sigma_i \sigma_j \tau_i \tau_j \tag{25}$$

In this case there are twelve possible damages per site which we label such that left means configuration A, right configuration B, top means σ and bottom means τ; for example $\left(\begin{smallmatrix} + & - \\ - & + \end{smallmatrix}\right)$ is $\sigma_i^A = +1$, $\tau_i^A = -1$, $\sigma_i^B = -1$ and $\tau_i^B = +1$. It can be shown [24] that if one fixes $\sigma_0^B = -1$ and $\tau_0^B = +1$ or $\sigma_0^B = +1$ and $\tau_0^B = -1$ then

$$\Gamma_i = \frac{\langle \sigma_i \tau_i \sigma_0 \tau_0 \rangle - \langle \sigma_i \tau_i \rangle \langle \sigma_0 \tau_0 \rangle}{2\,(1 - \langle \sigma_0 \tau_0 \rangle)} \tag{26}$$

with

$$\Gamma_i = \left(d_i^{\left(\begin{smallmatrix} + & + \\ + & - \end{smallmatrix}\right)} + d_i^{\left(\begin{smallmatrix} + & - \\ + & + \end{smallmatrix}\right)} + d_i^{\left(\begin{smallmatrix} - & - \\ - & + \end{smallmatrix}\right)} + d_i^{\left(\begin{smallmatrix} - & + \\ - & - \end{smallmatrix}\right)} \right) - \left(d_i^{\left(\begin{smallmatrix} + & + \\ - & + \end{smallmatrix}\right)} + d_i^{\left(\begin{smallmatrix} + & - \\ - & - \end{smallmatrix}\right)} + d_i^{\left(\begin{smallmatrix} - & + \\ + & + \end{smallmatrix}\right)} + d_i^{\left(\begin{smallmatrix} - & - \\ + & - \end{smallmatrix}\right)} \right) \tag{27}$$

and if one fixes $\sigma_0^B = -1$ then

$$\tilde{\Gamma}_i = \frac{\langle \sigma_i \sigma_0 \rangle - \langle \sigma_i \rangle \langle \sigma_0 \rangle}{2\left(1 - \langle \sigma_0 \rangle\right)} \tag{28}$$

with

$$\tilde{\Gamma}_i = \left(d_i^{\left(\substack{+-\\+-}\right)} + d_i^{\left(\substack{+-\\++}\right)} + d_i^{\left(\substack{+-\\-+}\right)} + d_i^{\left(\substack{+-\\--}\right)} \right) - \left(d_i^{\left(\substack{-+\\++}\right)} + d_i^{\left(\substack{-+\\+-}\right)} + d_i^{\left(\substack{-+\\-+}\right)} + d_i^{\left(\substack{-+\\--}\right)} \right) . \tag{29}$$

So, both types of correlation functions that one has in the Ashkin-Teller model can be expressed as a combination of damages.

The damage for which we have presented numerical data in the preceding section was not the quantity Γ_i but it was the sum of all the damages:

$$\Delta = \sum_i \Delta_i \quad \text{with} \quad \Delta_i = d_i^{+-} + d_i^{-+} \; . \tag{30}$$

In order to express these quantities in terms of thermodynamic quantities it is necessary to make some assumptions. Let us therefore restrict ourselves now to ferromagnetic interactions, heat bath dynamics and the use of the same random numbers for both configurations. We consider again only Ising variables and fix the damage as in Eq. (19), i.e. $\sigma_0^B = -1$. Since at $t = 0$ the damage is of type $+-$ at the origin we have

$$p_i^A \geq p_i^B \qquad \text{for all } i \tag{31}$$

where p_i^A is the value defined in Eq. (15a) for configuration A. Suppose one would try to create a damage of type $-+$ at site i. Then one would need, in order to produce $\sigma_i^A = -1$, a random number z which fulfills $z \geq p_i^A$ according to heat bath. Since one is using the same random number for configuration B this means using Eq. (31) that $z \geq p_i^B$ and therefore $\sigma_i^B = -1$. It is therefore impossible to create a damage of type $-+$ and therefore Eq. (31) will be valid also at the next time-step. By induction one can conclude now that Eq. (31) will always be valid and that a damage of type $-+$ cannot be created under the conditions that we had imposed. Consequently we have proved that $d^{-+} = 0$ and it follows for the damage

$$\Delta_i = \frac{C_{0i}}{2(1-m)} \quad \text{and} \quad \Delta = \frac{\chi}{2(1-m)} \tag{32}$$

where $\chi = \sum_i C_{0i}$ is the susceptibility.

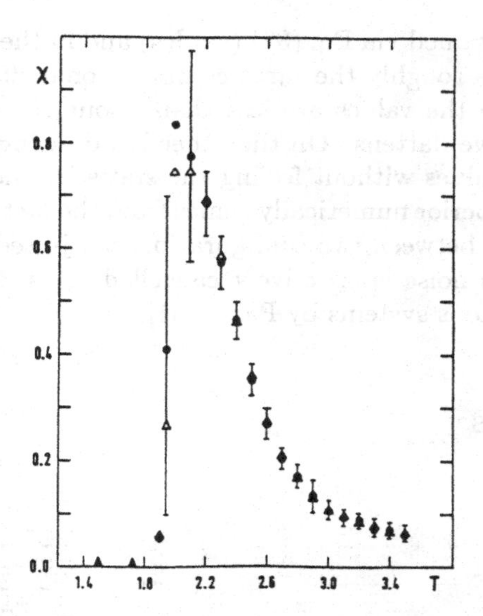

Figure 11. Susceptibility $\chi(\bullet)$ and $2\Delta(1-m)(\Delta)$ as function of temperature from 30 systems of size 10×10 (taken from Ref.29) for the 2d Ising model.

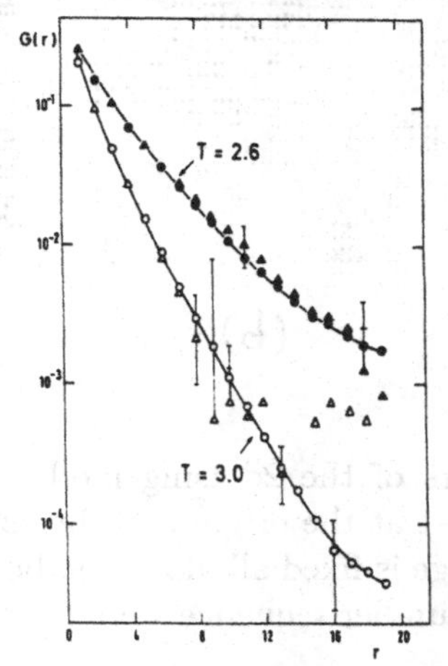

Figure 12. Correlation function $G(r) = \sum_{|i|=r} C_{0i}(\Delta$ for $T = 2.6$ and Δ for $T = 3.0)$ and $2\bar{\Delta}(r)(1-m)$ ($\bullet$ for $T = 2.6$ and $\circ$ for $T = 3.0$) where $\bar{\Delta} = \sum_{|i|=r} \Delta_i$ as a function of r in a semi-log plot. The data come from 10 systems of size 40×40 and were taken from Ref. 29 for a 2d Ising model.

In Figs. 11 and 12 we see how the two exact relations of Eq. (32) are realized numerically for the 2d Ising model. In Fig. 11 we see the susceptibility obtained from Eq. (32) on one hand and from the fluctuations of the magnetization on the other hand in a small system and the data agree very nicely. In Fig. 12 we see the correlation function obtained via Eq. (32) (circles) and in the usual way (triangles) using for both methods roughly the same computational effort. One sees that in the usual method once the values are less than about 10^{-3} the statistical noise takes over and the curve flattens. On the other hand, using Eq. (32) one gets to substantially smaller values without feeling the statistical noise. The reason why the use of damage is superior numerically comes from the fact that this method just monitors the difference between two configurations subjected to the same thermal fluctations so that this noise is effectively cancelled. A similar fact was already pointed out for continuous systems by Parisi [31].

2.6. DAMAGE CLUSTERS

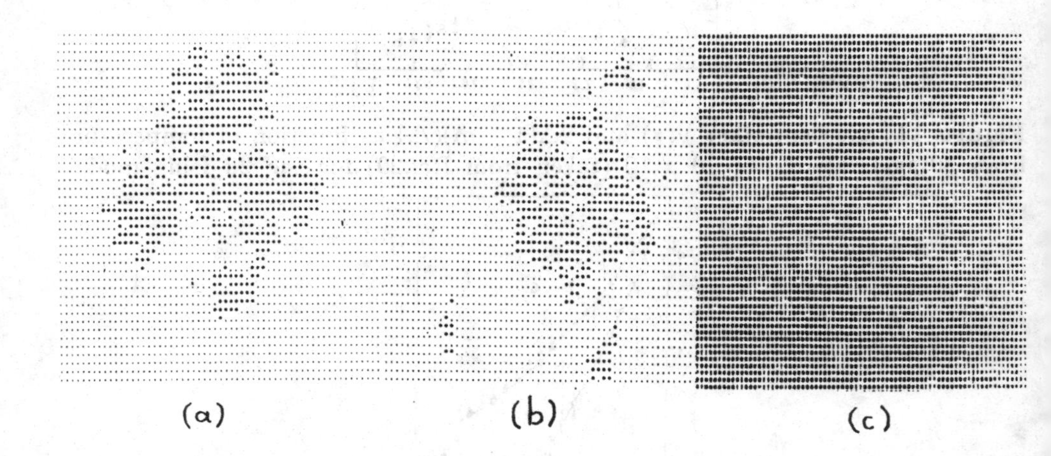

Figure 13. Damage clusters of the 2d Ising model at T_c. In (a) and (b) the damage is fixed to be $\sigma_0^B = -1$ at the origin and the size of the system is $L = 60$. In (c) the same kind of damage is fixed all along the boundary and $L = 100$. Heat bath and the same random number sequence were applied on both configurations (taken from Ref. 29).

The damage that was fixed at the origin at time $t = 0$ acts as a source from which constantly damage spreads away. At a given instant one can look at all the sites that are damaged and one will find a cloud or cluster of sites around the origin. These clusters fluctuate in size and shape and are not necessarily connected. In Figs. 13a and b we see two of these clusters for the two dimensional Ising model at T_c which are just 38 time steps apart. Since heat bath was used these clusters represent according to Eq. (32) the fluctuation of the magnetization due to the application of a local field at the origin. In Fig. 13c we see what happens if a damage $\sigma_i^B = -1$ is fixed all along the boundary of the system. Then clusters of the type shown in Figs. 13a and b overlap and one finds fluctuations that seem to be of all sizes. Using the equivalence of the Ising model to a lattice gas one can interprete these fluctuations as the fluctuations in local density which have actually been measured in a recent experiment [32].

Using Eq. (32) it is actually possible to calculate the dependence between radius and number of sites for the damage clusters. One finds that at the critical point the clusters are fractal, that means that if L is the size of the smallest box into which the cluster fits and s is the number of sites in the cluster (averaging over all clusters) one has a relation

$$s \propto L^{d_f} \tag{33}$$

Using Eq. (32) and the fact that at T_c one has $G(r) \propto r^{2-d-\eta}$ one obtains $d_f = d - \beta/\nu$ where β and ν are the critical exponents for the magnetization and for the correlation length.

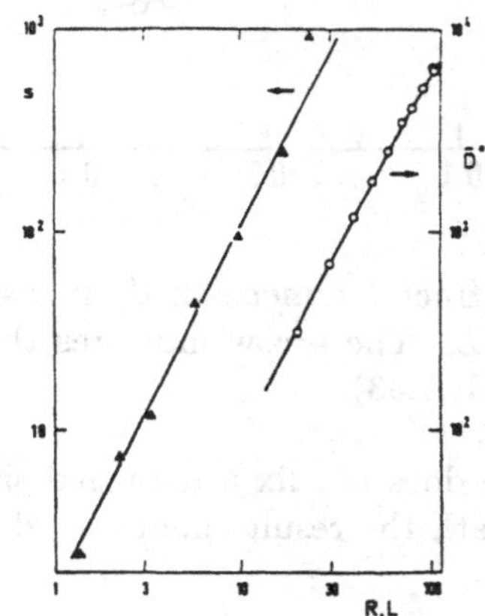

Figure 14. Log-log plot of the number s of damaged sites against the radius R of the cluster ($\triangle$) for the 2d Ising model at T_c with $\sigma_0^B = -1$ fixed. We also show the total number $\bar{D}^*$ of damaged sites if the damage is fixed on the boundary of a system of size L as a function of L (o) (taken from Ref. 29).

Since one does not expect several length scales in the problem one can also replace in Eq. (33) L by the radius of gyration R. In Fig. 14 we see results for the numerical determination of d_f for the two dimensional Ising model at T_c and the agreement with $d_f = \frac{7}{4}$ is reasonably good.

It is also possible to make similar arguments for the case when the damage is fixed on the boundary as shown in Fig. 13c. There the density of damaged sites in the center of the box decreases with a power law in the size of the box and consequently one finds again a fractal dimension as shown also in Fig. 14.

The fractal dimension of the clusters can also be measured by the touching method that has been widely used in cellular automata. One lets the damage spread until it touches the boundary of the system. The number of sites S of the touching cluster that are within a box of size ρ around the origin scales like $S \propto \rho^{d_{eff}}$ where d_{eff} is the effective fractal dimension which should converge with the system size L to the d_f of Eq. (33). Numerically it has been observed [33] that this convergence goes like $1/\ln L$ as seen in Fig. 15. The convergence using other methods can be even slower which might be due to multiscaling [33].

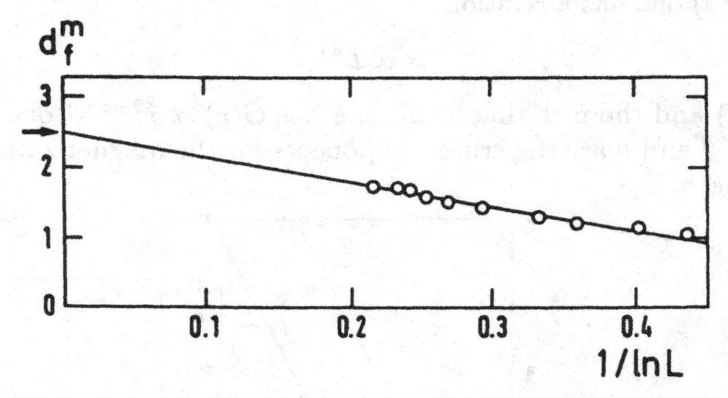

Figure 15. The effective fractal dimension d_f measured via the box counting method as function of $1/\ln L$. The arrow indicates the extrapolated value $d_f = d - \beta/\nu \simeq 2.5$ (taken from Ref. 33).

Let us note that if one does not fix a damaged site and if one uses Glauber dynamics instead of heat bath the result changes and one seems to find compact clusters [25].

2.7. DAMAGE IN SPIN GLASSES

Numerical work on spin glasses has been challenging and frustrating in the past so that any new method that gives some hope of improving the results should be tested. The results obtained for the $\pm$ Ising spin glass of damage spreading using heat bath dynamics [28] are shown in Fig. 16. In their spirit they are analogous to the data of the pure Ising model shown in Fig. 10. In three dimensions one believes that about $T_{SG} = 1.2$ there is a transition to a spin glass phase. At $T_G = 4.5$, the critical temperature of the pure Ising model, one believes that the paramagnetic phase changes into a so-called Griffiths phase in which correlations decay a little shower than exponential; but this phase is very difficult to detect or do discern from the paramagnetic behaviour.

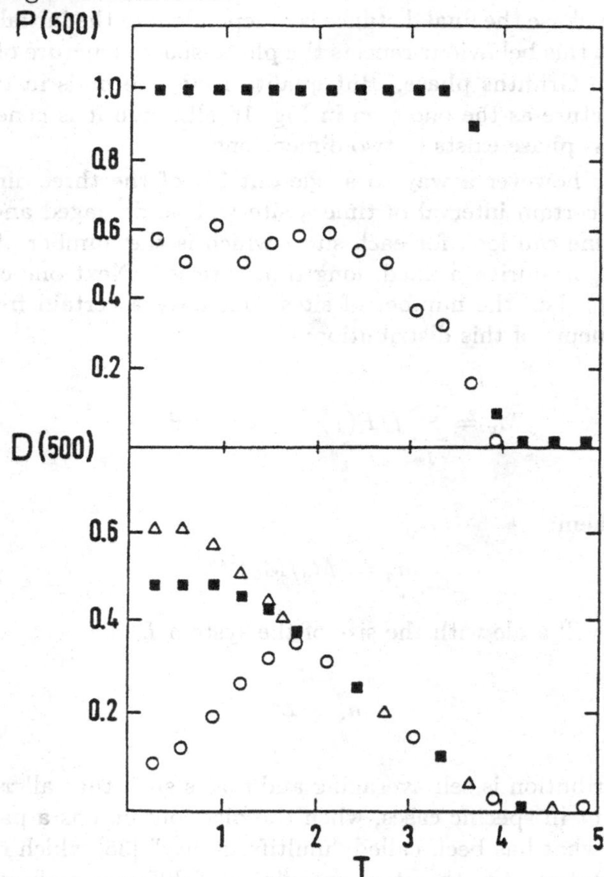

Figure 16. Damage in the three dimensional ± spin glass after 500 time steps as a function of temperature for $L = 12$ taken from Ref. 28 for various initial damages: $\Delta(0) \longrightarrow 0$ (○), $\Delta(0) = \frac{1}{2}$ (□) and $\Delta(0) = 1$ (△); (a) fraction $P(t)$ of non-identical configurations and (b) distance $D(t)$ between two configuration provided they are still different.

¿From Fig. 16 one sees indeed two characteristic temperatures not too far from T_{SG} and T_G which could be interpreted as separating three phases. As opposed to the ferromagnetic phase of Fig. 10 both low temperature phases here are chaotic because the probability of two configurations to be still different after some time is not zero for times large but much shorter than the Poincaré time. In the low temperature phase the final distance does depend on the initial distance while in the intermediate phase the final distance is independent of the initial distance. One might argue that this behaviour reflects the phase-space structure of the spin-glass phase and of the Griffiths phase. But qualitatively one finds in two dimensions [34] the same picture as the one seen in Fig. 16 although it is generally accepted that no spin glass phase exists in two dimensions.

There exists however a way to single out T_G of the three-dimensional spin glass. During a certain interval of time a site will be damaged and healed again several times. One can look for each site i which is the number of times f_i that it is damaged again during a fixed, long time interval. Next one can look at the distribution $P(f)$, i.e. the number of sites that have a certain frequency f and analyze the moments of this distribution:

$$M_q = \sum_f f^q P(f) \qquad q = 0, \pm 1, \cdots \tag{34}$$

and normalize them:

$$m_q = (M_q/M_0)^{1/q} \tag{35}$$

These moments will scale with the size of the system L like

$$m_q \propto L^{x_q} \tag{36}$$

Usually the distribution is self averaging and scales such that all x_q (with $q \neq 0$) are the same. But in specific cases, when the distribution has a particularly long tail it can show what has been called "multifractality" [35] which means that the x_q vary with q and one has therefore an infinity of different scaling exponents.

In Ref.36 the moments of Eq.(35) have been calculated for various models. For the three dimensional spin glass it has been found that the distribution is multifractal at T_G as can be seen from Fig. 17a. This is in marked contrast to what has been found at T_{SG} for the 3d spin glass, at T_G or T_{SG} for the 2d spin glass or at the critical temperature T_c of the 3d Ising model because in all these cases the lines for the different moments in the log-log plot are parallel to each other which means that x_q is the same for all q. As an example we show in Fig. 17b the data for the 3d Ising model.

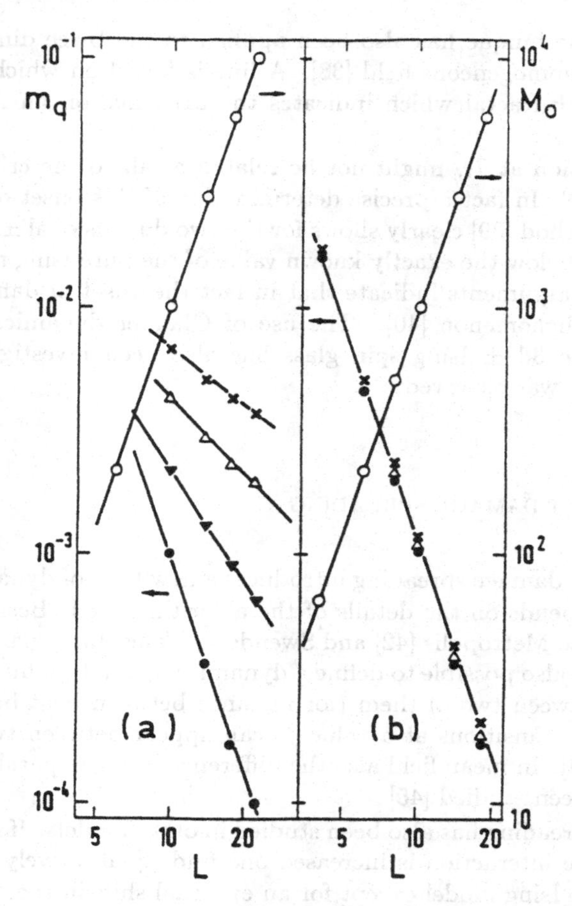

Figure 17. Log-log plot of the moments of the distribution $P(f)$ of the redamaging of sites as a function of the system size L for (a) the 3d $\pm$ spin glass at T_G (b) the standard 3d Ising model at T_c. The moments are: $M_0(\circ)$, $m_1(\bullet)$, m_2 (∇), m_3 (Δ) and $m_4(\times)$ (taken from Ref.36).

We can conclude that the temperature $T_G = 4.5$ of the 3d $\pm$ spin glass is particular because there the distribution of the redamaging frequency is multifractal while in the usual case, particularly at T_G in the 2d $\pm$ spin glass one has a simple one-exponent scaling behaviour. This feature might be an empirical way to single out models that show spin glass behaviour. The numerical evidence that T_{SG} is indeed the transition to the spin glass phase can be strengthened by making a scaling Ansatz for the relaxation times [37].

The same technique has also been applied to the three dimensional ± Ising spin glass in a homogeneous field [38]. A line is found on which the redamaging frequency is multifractal which indicates the existence on an Almeida-Thouless line also in 3d.

The transition at T_G might not be related at all to the critical point of the pure Ising model. In fact a precise determination of this onset of spreading using the gradient method [39] clearly shows for the two dimensional ± Ising spin glass a value of T_G 6% below the exactly known value of the pure Ising model. Numerical and theoretical arguments indicate that in fact the onset of damage is related to a percolation phenomenon [40]. The use of Glauber dynamics for the damage spreading in the 3d ± Ising spin glass has also been investigated [41] and no transition at T_G was observed.

2.8. MORE ABOUT DAMAGE SPREADING

Apparently damage spreading introduces a new type of dynamical phase transition which depends on the details of the algorithm used. Besides for heat bath and Glauber also Metropolis [42] and Swendsen-Wang [43] dynamics have been investigated. It is also possible to define a dynamics with a tunable parameter f that interpolates between two of them (for instance between heat bath and Glauber) and then phase transitions at a value f_c can appear between two different types of behaviour [44]. In mean field also the difference between parallel and sequential dynamics has been studied [45]

Damage spreading has also been studied in other models. If in the Ising model the range of the interaction is increased one finds qualitatively the same picture as in the simple Ising model except for an eventual shift in the onset of spreading temperature [46]. In the ANNNI model it seems that damage spreading constitutes a way to observe the floating phase in two dimensions [47]. Continuous degrees of freedom have also been considered and as well for the XY-model [23] as for the Heisenberg model [48] unexpected phase transitions have been found.

Acknowledgments

I thank D. Bessis, A. Coniglio, L. de Arcangelis, G. Mantica and A. Mariz with whom I collaborated on this subject and L. Knopoff and K. R. Sreenivasan for useful discussions.

References

[1] McCann, W., S. Nishenko, L. Sykes, J. Krause, Pageoph **117**, 1082 (1979); C. Lomnitz, Bull. Seism. Soc. Am. **72**, 1441 (1982).

[2] Sammis, C., G. King, R. Biegel, Pageoph **125**, 777 (1987).

[3] Mandelbrot, B.B., *The Fractal Geometry of Nature*, Freeman, San Francisco, 1982.

[4] Boyd, D.W., Mathematica **20**, 170 (1973); see also ref. 3.

[5] Falconer, J.K., *The Geometry of Fractal Sets*, Cambridge Univ. Press, 1985; L.R. Ford, *Automorphic Functions*, Chelsen Publ., 1929.

[6] Herrmann, H.J., G. Mantica, D. Bessis, Phys. Rev. Lett. **65**, 3223 (1990).

[7] Manna, S.S., H.J. Herrmann, preprint.

[8] for a review see G. Paladin and A. Vulpiani, Phys. Rep. **156**, 147 (1987) or A. Coniglio, L. de Arcangelis and H.J. Herrmann, Physica A **157**, 21 (1989).

[9] Huber, G., private communication.

[10] Manna, S.S., T. Vicsek, preprint.

[11] Bessis, D., S. Demko, Comm. in Math. Phys., preprint.

[12] Batchelor, G.K., *Theory of homogeneous turbulence*, Cambridge Univ. Press, 1982.

[13] Kauffman, S.A., J. Theor. Biol. **22**, 437 (1969).

[14] Derrida, B., Y. Pomeau, Europhys Lett. **1**, 45 (1986); B. Derrida, D. Stauffer, Europhys Lett. **2**, 739 (1986).

[15] da Silva, L.R., H.J. Herrmann, J. Stat. Phys. **52**, 463 (1988).

[16] Wolfram, S., *Theory and Application of Cellular Automata*, World Scientific, Singapore, 1986; E. Bienenstock, F. Fogelman-Soulié, G. Weisbuch (eds), *Disordered Systems and Biological Organization*, Springer, NATO ASI Series F, 1986.

[17] Herrmann, H.J., in *Nonlinear Phenomena in Complex Systems*, A.N. Proto (ed), Elsevier, Amsterdam, 1989.

[18] Pomeau, Y., J. Phys. A **17**, L415 (1984); H.J. Herrmann, J. Stat. Phys. **45**, 145 (1986).

[19] Zabolitzky, J.G., H.J. Herrmann, J. Comp. Phys. **76**, 426 (1988); S.C. Glotzer, D. Stauffer, S. Sastry, Physica A **164**, 1 (1990).

[20] Stauffer, D., in *Computer Simulation Studies in Condensed Matter Physics II*, D.P. Landau, K.K. Mon and H.-B. Schüttler (eds), Springer, Heidelberg, 1990.

[21] de Arcangelis, L., A. Coniglio, Europhys Lett. **7**, 113 (1988).

[22] Hilhorst, H.J., M. Nijmeijer, J. Physique **48**, 185 (1987).

[23] Golinelli, O., B. Derrida, J. Phys. A **22**, L939 (1989).

[24] Mariz, A.M., J. Phys. A **23**, 979 (1990).

[25] Stanley, H.E., D. Stauffer, J. Kertész, H.J. Herrmann, Phys. Rev. Lett. **59**, 2326 (1987).
[26] Costa, U., J. Phys. A **20**, L583 (1987).
[27] Le Caër, G., J. Phys.A **22**, L647 (1989) and Physica A **159**, 329 (1989).
[28] Derrida, B., G. Weisbuch, Europhys Lett. **4.**, 657 (1987).
[29] Coniglio, A., L. de Arcangelis, H.J. Herrmann, N. Jan, Europhys. Lett. **8**, 315 (1989).
[30] Ashkin, J., E. Teller, Phys. Rev. **64**, 178 (1943).
[31] Parisi, G., Nucl. Phys. B **180**, 378 (1981).
[32] Guenoun, P., F. Perrot, D. Beysens, Phys. Rev. Lett. **63**, 1152 (1989).
[33] de Arcangelis, L., H.J. Herrmann, A. Coniglio, J. Phys. A **23**, L265 (1990).
[34] Neumann, A.U., B. Derrida, J. Physique **49**, 1647 (1988).
[35] see e.g. G. Paladin and A. Vulpiani, Phys. Rep. **156**, 147 (1987).
[36] de Arcangelis, L., A. Coniglio, H.J. Herrmann, Europhys Lett. **9**, 749 (1989).
[37] Campbell, I.A., L. de Arcangelis, Europhys Lett. ...
[38] de Arcangelis, L., H.J. Herrmann, A. Coniglio, J. Phys. A **22**, 4659 (1989).
[39] Boissin, N., H.J. Herrmann, J. Phys. A **24**, L43 (91).
[40] de Arcangelis, L., A. Coniglio, preprint; A. Coniglio, in *Correlations and Connectivity in Constrained Geometries*, H.E. Stanley and N. Ostrowsky (eds), Kluwer, Dordrecht, 1990.
[41] da Cruz, H.R., U.M.S. Costa and E.M.F Curado, J. Phys. A **22**, L651 (1989).
[42] Mariz, A.M., H.J. Herrmann, L. de Arcangelis, J. Stat. Phys. **59**, 1043 (1990).
[43] Stauffer, D., Physica A **162**, 27 (1990).
[44] Mariz, A.M., H.J. Herrmann, J. Phys. A **22**, L1081 (1989); O. Gollinelli, Physica A **167**, 736 (1990).
[45] Golinelli, O., B. Derrida, J. Physique **49**, 1663 (1988).
[46] Manna, S.S., J. Physique **51**, 1261 (1990).
[47] Barber, M.N., B. Derrida, J. Stat. Phys. **51**, 877 (1988).
[48] Miranda, E., preprint HLRZ 82/90.

A BRIEF ACCOUNT OF STATISTICAL THEORIES OF LEARNING AND GENERALIZATION IN NEURAL NETWORKS

P. PERETTO
M. GORDON
M. RODRIGUEZ-GIRONES
C.E.N. Grenoble. DRF/SPh.
BP85X 38041 Grenoble
France

1. Introduction

A network of automata is made up of distinct units (or automata) i ; $i = 1, \cdots, N$. One assumes that an automaton may be in one of a (generally finite) number of internal states σ_i. At (discrete) time ν the overall state $I(\nu)$ of the network is defined as the set of states that the units take at this very time : $I(\nu) \equiv \{\sigma_i(\nu)\}$. The time evolution of the network is driven by a dynamics which depends on a number of parameters $\{J\}$. The parameters $\{J\}$, which determine the structure of the network, define how the units influence each other and how information proceeds through the system.

Programming a network consists in choosing the parameters $\{J\}$ so as to make the fixed point of the dynamics $I(\nu \to \infty)$, the solution of the problem. For example classical computers solve problems by modifying $I = \{\sigma\}$, the set of its states, called the data, according to a dynamics which is determined by the program that is to say by the set of parameters $\{J\}$. At the very beginning of automatic calculation data and programs were considered as quite different entities. We have Von Neumann to thank for having realized that both pieces of information could be treated on equal footing and stored in the same memory. This conceptual jump has been at the root of the development of modern computers. Similarly one can think that the state of more general automata networks is described by $\{\sigma\} \otimes \{J\}$, that is to say by the set of internal states and the set of network parameters. Usually the parameters $\{J\}$, the program, are given once and for all at the beginning of the solving session. The states $\{\sigma\}$ are the sole dynamical variables. It could be interesting to allow the parameters $\{J\}$ to also be dynamical variables. Then one says that the system is endowed with *learning capabilities*. "Plastic networks" is a class of automata networks which is the focus of increasing attention from computer scientists.

E. Goles and S. Martínez (eds.), Statistical Physics, Automata Networks and Dynamical Systems, 119–171.

These efforts, however, concern deterministic machines only. It happens that one domain where automata networks prove to be very useful is the study the properties of physical systems. As physical systems are noisy the automata are necessarily probabilistic : Given $I(\nu - 1)$ the state of the network at time $\nu - 1$, its state at time ν is known only through a probability distribution $\rho(I, \nu)$. Far from introducing unwanted complications, stochasticity naturally allows paths towards putative solutions of a problem to be analysed in parallel, even for standard questions as those arising in combinatorial optimization problems for example. This is the reason why this paper deals with probabilistic automata networks. Now, how probabilistic automata may learn is a challenging question. It is one of the topics of this paper.

The article is organized along the following lines. The natural framework for the description of probabilistic networks is that of statistical mechanics. This is the subject of section 2. How these concepts may be applied to the problems of learning and generalization is considered in section 3. An application to specific systems, namely perceptrons, is treated in the following section, section 4, and, finally, one gives some insight on the difficult problem of learning and generalizing in more general networks in section 5.

2. Statistical Mechanics of Automata Networks

2.1. ENSEMBLES OF NETWORKS

A theory of automata networks may be built by using the tools of statistical mechanics. Physicists are interested in studying automata networks because these machines can model most of natural systems. However modelling involves some simplification, in particular the neglect of dynamical variables which are considered as being of secundary importance. The influence of these hidden dynamical variables is summed out. They manifest themselves by making the dynamics non deterministic, that is to say the successor state $\mathcal{I}(\nu + 1)$ of a given state $\mathcal{I}(\nu)$ is not known for sure but only with a certain probability. In this section $\mathcal{I}$ may be $\{\sigma\}$ or $\{J\}$ or a combination of both sets of dynamical variables. The set of hidden variables is called the reservoir of the system and the state of the reservoir is generally determined by a single parameter, the noise parameter (also called the *temperature* of the reservoir). Due to the probabilistic character of the dynamics, a statistical description is necessary and one introduces *ensembles* of networks. An ensemble is a set of systems which share some common properties. For example a set of machines, all comprised of N binary threshold units, makes an ensemble of neural networks of equal sizes. An element of the ensemble is characterized by its state $\mathcal{I}$ and, counting the relative number of elements which are in state $\mathcal{I}$ at time ν, defines a probability distribution $\rho(\mathcal{I},\nu)$.

One now assumes that the dynamics is Markovian. This amounts to saying that the dynamics is fully determined by a matrix $\mathbf{W}$ whose elements $W(\mathcal{I} \mid \mathcal{J})$ represent the probability for a network to be in state $\mathcal{I}$ at time $\nu+1$ given it is in state $\mathcal{J}$ at time ν. The evolution of the probability distribution is then driven by the following *master equation* :

$$\rho(\mathcal{I},\nu+1) = \sum_{\mathcal{J}} W(\mathcal{I} \mid \mathcal{J})\rho(\mathcal{J},\nu) \tag{1}$$

or, using a matrix notation :

$$\tilde{\rho}(\nu+1) = \mathbf{W}.\tilde{\rho}(\nu) \quad ; \quad \tilde{\rho}(\nu) \equiv \{\rho(\mathcal{I},\nu)\}$$

with:

$$\sum_{\mathcal{I}} W(\mathcal{I} \mid \mathcal{J}) = 1 \quad \text{and} \quad \sum_{\mathcal{I}} \rho(\mathcal{I},\nu) = 1 \tag{2}$$

The equations (1) and (2) reduce to deterministic dynamics if the successor state $\mathcal{I}$ of a state $\mathcal{J}$ is known for sure, that is to say if $W(\mathcal{I} \mid \mathcal{J}) = 1$ and $W(\mathcal{I}' \mid \mathcal{J}) = 0$ for $\mathcal{I}' \neq \mathcal{I}$. If one assumes, moreover, that the initial distribution is peaked on a particular state $\mathcal{I}^0$, the equations (1) and (2) describe the evolution of a particular deterministic machine. The statistical framework therefore encompasses all sorts of dynamics, be they stochastic or deterministic.

In the latter case at least, the stochastic approach seems to introduce, at first sight, unnecessary mathematical complexity. However, as it often happens in mathematical physics, changing a class of problems into a larger one, makes the problems amenable to analytical treatment. If one is really interested in the properties of deterministic networks it is then possible to consider the results obtained by studying the probabilistic systems and by looking how the solutions evolve when reducing the noise parameter to zero.

The steady distribution, $\tilde{\rho}(\nu \to \infty)$, $\tilde{\rho}$ for short, is of particular interest. Due to the special properties of statistical matrices which are displayed in equation (2), this distribution is the right eigenvector of $\mathbf{W}$ for the (Perron) eigenvalue $\lambda = 1$:

$$\mathbf{W}.\tilde{\rho} = \tilde{\rho} \tag{3}$$

For example $\rho(\mathcal{I})$ is the probability of finding a network in state $\mathcal{I}$ among a steady population of networks belonging to the same ensemble. The power of statistical mechanics comes from the fact that it is possible to find the steady distribution

$\tilde{\rho}$ for a large class of systems without explicitely solving the complicated equation (3) as explained in the following section.

2.2. INTRODUCING COST FUNCTIONS

Let us associate a scalar field $H(\mathcal{I})$ to the phase space $\mathcal{I}$ of the ensemble of networks. This scalar is called a cost. This means that it "costs" $H(\mathcal{I})$ to have a network in state $\mathcal{I}$. How the cost function is determined depends on the problem the network has to solve and also on the ingenuity of the modeler. We shall see examples below.

One assumes that :

a) the steady distribution is fully determined by the cost :

$$\rho(\mathcal{I}) = \Phi(H(\mathcal{I})) \tag{4}$$

with the higher the cost the lower the probability for the system to be in state $\mathcal{I}$,

b) the cost is an additive function : It is possible to split the problem into two sub-problems a and b or the machine into two sub-machines a and b with respective costs H_a and H_b and probabilities $\rho(H_a)$ and $\rho(H_b)$ such that the cost for the whole problem is the sum of partial costs :

$$H = H_a + H_b$$

(that is to say the cost brought about by the interaction H_{ab} between the two parts can be neglected) and the probability for the whole problem is the product of probabilities.

The solution of the equation :

$$\rho(H) \equiv \rho(H_a + H_b) = \rho(H_a)\rho(H_b)$$

is given by :

$$\rho(H) = \frac{1}{Z} \exp -\frac{H}{T} \tag{5}$$

Z is a normalization constant :

$$Z = \sum_{\mathcal{I} \in \mathcal{I}^H} \exp -\frac{H(\mathcal{I})}{T} \tag{6}$$

where $\mathcal{I}^H$ is the set of states $\mathcal{I}$ for which a cost $H(\mathcal{I})$ has been defined. The distribution $\rho(H)$ given by eq. (5) is called the Maxwell-Boltzmann distribution.

Z is called the *partition function.* In the present approach $\mathcal{I}^H$ does not necessarily span the whole phase space of $\mathcal{I}$. This may look a bit strange but it is not actually if one realizes that this formulation is equivalent to assuming that the cost for the states $\mathcal{I}$ not belonging to $\mathcal{I}^H$ is very large, so making all contributions of these states to the partition function vanish :

$$\begin{aligned} H(\mathcal{I}) &\quad \text{for } \mathcal{I} \in \mathcal{I}^H \\ H(\mathcal{I}) \to \infty &\quad \text{for } \mathcal{I} \notin \mathcal{I}^H \end{aligned}$$

Then the classical definition of Z is recovered :

$$Z = \sum_{\text{all } \mathcal{I}} \exp - \frac{H(\mathcal{I})}{T} \tag{6'}$$

Let O be a property of the system (an observable) we are interested in. $O(\mathcal{I})$ is the value of the property when the network is in state $\mathcal{I}$. The statistical average at time ν of O is given by :

$$\langle O(\nu) \rangle = \sum_{\mathcal{I}} \rho(\mathcal{I}, \nu) O(\mathcal{I}) \tag{7}$$

At statistical equilibrium, the average becomes :

$$\langle O \rangle = \sum_{\mathcal{I}} \rho(\mathcal{I}) O(\mathcal{I}) \tag{8}$$

where the sum is over all states $\mathcal{I} \in \mathcal{I}^H$.

Two observables are of particular interest, the cost and the entropy. According to eq. (8) the average cost U is given by :

$$U = \langle H \rangle = \sum_{\mathcal{I}} \rho(\mathcal{I}) H(\mathcal{I}) \tag{9}$$

The entropy S is defined as the average of the logarithm of the distribution probability :

$$S = - \langle \mathrm{Log}(\tilde{\rho}) \rangle = - \sum_{\mathcal{I}} \rho(\mathcal{I}) \mathrm{Log}\left[\rho(\mathcal{I})\right] \tag{10}$$

Since $0 \leq \rho(\mathcal{I}) \leq 1$, the entropy S is necessarily a non negative quantity. A linear combination of U and S, $F = U - TS$, plays a central role in statistical mechanics.

F is called the free energy of the system. Using the definitions (9) and (10) it is written as :

$$F = \sum_{\mathcal{I}} \rho(\mathcal{I}) H(\mathcal{I}) + T \sum_{\mathcal{I}} \rho(\mathcal{I}) \mathrm{Log}\,[\rho(\mathcal{I})] \tag{11}$$

Introducing the explicit expression of $\rho(\mathcal{I})$

$$\rho(\mathcal{I}) = \frac{1}{Z} \exp\left[-\frac{H(\mathcal{I})}{T}\right]$$

one obtains :

$$\begin{aligned} F &= \sum_{\mathcal{I}} \rho(\mathcal{I}) H(\mathcal{I}) + T \sum_{\mathcal{I}} \rho(\mathcal{I}) \left[-\frac{H(\mathcal{I})}{T} - \mathrm{Log}(Z)\right] \\ &= \sum_{\mathcal{I}} \rho(\mathcal{I}) H(\mathcal{I}) - \sum \rho(\mathcal{I}) H(\mathcal{I}) - T \sum_{\mathcal{I}} \rho(\mathcal{I}) \mathrm{Log}(Z) \\ &= -\,T \mathrm{Log}(Z) \end{aligned} \tag{12}$$

where the normalization of the distribution has been used (cf. eq. (2)). The free energy F is simply given by the logarithm of the partition function Z.

Its real interest comes from the property that F is minimal for all perturbations which modify the distribution. To prove this statement we consider two distributions, the Boltzmann distribution $\rho(\mathcal{I})$ and another distribution $\rho'(\mathcal{I})$. In both distributions $\mathcal{I} \in \mathcal{I}^H$. According to eq. (11) the free energy associated with ρ' is given by :

$$F' = \sum_{\mathcal{I}} \rho'(\mathcal{I}) H(\mathcal{I}) + T \sum_{\mathcal{I}} \rho'(\mathcal{I}) \mathrm{Log}\,[\rho'(\mathcal{I})]$$

and the free energy difference by :

$$\begin{aligned} \Delta F &= F' - F \\ &= \sum_{\mathcal{I}} [\rho'(\mathcal{I}) - \rho(\mathcal{I})]\, H(\mathcal{I}) + T \sum_{\mathcal{I}} [\rho'(\mathcal{I}) \mathrm{Log}(\rho'(\mathcal{I})) - \rho(\mathcal{I}) \mathrm{Log}(\rho(\mathcal{I}))] \end{aligned} \tag{13}$$

Using

$$\mathrm{Log}(\rho'(\mathcal{I})) = \mathrm{Log}\left[\frac{\rho'(\mathcal{I})}{\rho(\mathcal{I})} \rho(\mathcal{I})\right] = \mathrm{Log}\left[\frac{\rho'(\mathcal{I})}{\rho(\mathcal{I})}\right] + \mathrm{Log}(\rho(\mathcal{I}))$$

the last term of eq. (13) is written as :

$$\rho'(\mathcal{I})\text{Log}\left[\frac{\rho'(\mathcal{I})}{\rho(\mathcal{I})}\right] + \rho'(\mathcal{I})\text{Log}(\rho(\mathcal{I})) - \rho(\mathcal{I})\text{Log}(\rho(\mathcal{I})) = \cdots$$

$$\cdots = \rho'(\mathcal{I})\text{Log}\left[\frac{\rho'(\mathcal{I})}{\rho(\mathcal{I})}\right] + [\rho'(\mathcal{I}) - \rho(\mathcal{I})]\,\text{Log}(\rho(\mathcal{I}))$$

Therefore

$$\Delta F = \sum_{\mathcal{I}} [\rho'(\mathcal{I}) - \rho(\mathcal{I})]\,[H(\mathcal{I}) + T\text{Log}(\rho(\mathcal{I}))] + T\sum_{\mathcal{I}} \rho'(\mathcal{I})\text{Log}\left[\frac{\rho'(\mathcal{I})}{\rho(\mathcal{I})}\right]$$

Taking the analytical expression of the Boltzmann distribution given in eqs. (5) and (6) into account, we have :

$$H(\mathcal{I}) + T\text{Log}(\rho(\mathcal{I}) = H(\mathcal{I}) + T\left[-\frac{H(\mathcal{I})}{T} - \text{Log}(Z)\right] = -T\text{Log}(Z)$$

and

$$\sum_{\mathcal{I}} [\rho'(\mathcal{I}) - \rho(\mathcal{I})]\,[H(\mathcal{I}) + T\text{Log}(\rho(\mathcal{I}))] = -T\text{Log}(Z)\sum_{\mathcal{I}} [\rho'(\mathcal{I}) - \rho(\mathcal{I})]$$

$$= -T\text{Log}(Z)\,[1-1] = 0$$

whence :

$$\Delta F = T\sum_{\mathcal{I}\in\mathcal{I}^H} \rho'(\mathcal{I})\text{Log}\left[\frac{\rho'(\mathcal{I})}{\rho(\mathcal{I})}\right]$$

The modification ΔF of the free energy may be written as :

$$\Delta F = -T\sum_{\mathcal{I}} \rho'(\mathcal{I})\text{Log}\left[\frac{\rho(\mathcal{I})}{\rho'(\mathcal{I})}\right]$$

One then notes that

$$\text{Log}(x) \le x - 1 \quad \forall\, x$$

(Fig. 1) where the equality holds for and only for $x = 1$.

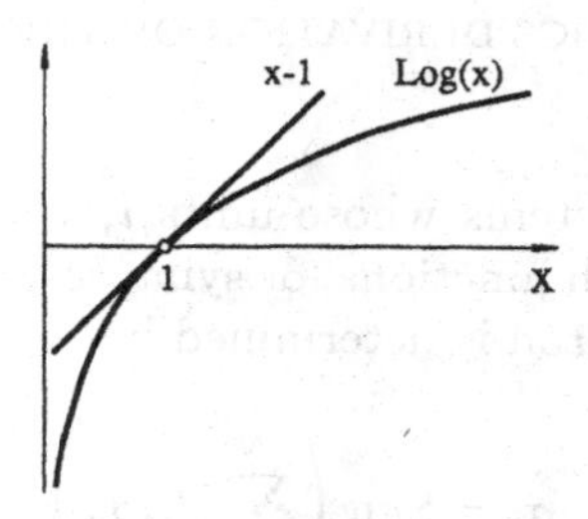

Figure 1

Therefore

$$\sum_{\mathcal{I}} \rho'(\mathcal{I})\text{Log}\left[\frac{\rho(\mathcal{I})}{\rho'(\mathcal{I})}\right] \leq \sum_{\mathcal{I}} \rho'(\mathcal{I})\left[\frac{\rho(\mathcal{I})}{\rho'(\mathcal{I})} - 1\right] = \sum_{\mathcal{I}} [\rho(\mathcal{I}) - \rho'(\mathcal{I})] = 0 \quad (14)$$

Since

$$\sum_{\mathcal{I}} \rho'(\mathcal{I})\text{Log}\left[\frac{\rho(\mathcal{I})}{\rho'(\mathcal{I})}\right] \leq 0$$

one concludes that

$$\Delta F \geq 0 \quad (15)$$

ΔF vanishes if and only if the two distributions are identical $\rho'(\mathcal{I}) \equiv \rho(\mathcal{I}) \ \forall \mathcal{I}$. The proof that it is necessary that the two distributions have to be identical for the equality to hold in eq. (15) rests upon the fact that if there exists a single state $\mathcal{I}$ such as $\rho'(\mathcal{I}) \neq \rho(\mathcal{I})$ the inequality in eq. (14) has to be strict. Equation (15) summarizes the fundamental H-theorem of Boltzmann.

The analytical expression eq. (14) giving the variation of the free energy brought about by the modification of the Boltzmann distribution $\rho(\mathcal{I})$ into a new distribution $\rho'(\mathcal{I})$ is also extremely interesting. It shows that the *increase of the free energy is given by the Kullback distance between the two distributions.* The Kullback distance between two distributions $\rho_1(\mathcal{I})$ and $\rho_2(\mathcal{I})$, which is also called the *relative entropy* between the two distributions, is defined by :

$$\mathcal{S} = \sum_{\mathcal{I}} \rho_1(\mathcal{I})\text{Log}\left[\frac{\rho_1(\mathcal{I})}{\rho_2(\mathcal{I})}\right]$$

This is a non-negative quantity which vanishes when and only when the distributions ρ_1 and ρ_2 are one another identical. Conversely one can consider the relative entropy between two distributions as an energy, a cost that some learning mechanism could strive to minimize.

2.3. A STATISTICAL MECHANICS DERIVATION OF THE DYNAMICS OF NEURAL NETWORKS

Neural networks are systems whose units i, $i = 1, \cdots, N$, the neurons, are connected each other through junctions, or synapses, whose strengths are J_{ij}. The state $\sigma_i \in \{-1, +1\}$ of a neuron is determined by :

$$\sigma_i = \text{Sgn}\left(\sum_{j=0}^{N} J_{ij}\sigma_j\right) \quad (16)$$

The neuron $j = 0$ is special. Its state is fixed to $\sigma_0 = -1$ and the connexion J_{i0} is called the threshold of neuron i. The sum :

$$h_i = \sum_{j=0}^{N} J_{ij}\sigma_j \tag{17}$$

is called the (local) field acting upon site i. To make sure that the dynamics of the network strives to fulfil the set of conditions given by eq. (16) one introduces cost functions. Two types of cost can be imagined :

a) On the one hand one may say that a cost is attached to the situation where

$$\sigma_i = -\text{Sgn}(h_i)$$

since it violates the rule (16) of neural networks. Counting the number of such cases when the system is in state $I = \{\sigma_i\}$ leads to a definition of cost based upon *errors counting.* The cost is then given by :

$$H(\mathcal{I}) = \sum_{i=1}^{N} \theta\left[-\sigma_i(I)h_i(\mathcal{I})\right] \quad ; \mathcal{I} \equiv I \otimes \{J\} \tag{18}$$

where the step function $\theta(x)$ is defined by :

$$\theta(x) = \begin{cases} 1 & \text{if } x > 0 \\ 0 & \text{if } x \leq 0 \end{cases}$$

The quantity :

$$\gamma_i = \sigma_i h_i = \sum_{j=0}^{N} J_{ij}\sigma_i\sigma_j \tag{19}$$

is of special interest. It is a measure of how stable is the state σ_i. It is called the *stabilization parameter* on site i.

$$H(\mathcal{I}) = \sum_{i=1}^{N} \theta\left[-\gamma_i(\mathcal{I})\right]$$

b) In equation (18) all errors are treated on equal footing. An error is an error. However some are more serious than others : Large negative stabilization parameters are associated with serious errors and small negative stabilization parameters with slighter ones. It seems quite natural to attach a heavier cost to the former than to the latter. This leads to a new definition of the cost function which is based upon ***stabilization intensity*** :

$$H(\mathcal{I}) = -\sum_{i=1}^{N} \gamma_i(\mathcal{I}) \tag{20}$$

The time evolution of a Markovian system is fully determined by the transition probabilities $W(\mathcal{I} \mid \mathcal{J})$ that the system is in state $\mathcal{I}$ at time $\nu + 1$ given it is in state $\mathcal{J}$ at time ν. For the asymptotic distribution to be the Maxwell-Boltzmann distribution it enough (and it is necessary) that these matrix elements obey the ***detailed balance principle*** :

$$W(\mathcal{I} \mid \mathcal{J})\text{exp}\left[\frac{H(\mathcal{I})}{T}\right] = W(\mathcal{J} \mid \mathcal{I})\text{exp}\left[\frac{H(\mathcal{J})}{T}\right] \tag{21}$$

Then

$$\rho(\mathcal{I}, \nu \to \infty) = \frac{1}{Z}\text{exp}\left[-\frac{H(\mathcal{I})}{T}\right]$$

We prove this statement by computing the following expression :

$$\begin{aligned}\sum_{\mathcal{J}} W(\mathcal{I} \mid \mathcal{J})\text{exp}\left[-\frac{H(\mathcal{J})}{T}\right] &= \sum_{\mathcal{J}} W(\mathcal{J} \mid \mathcal{I})\text{exp}\left[\frac{H(\mathcal{J}) - H(\mathcal{I})}{T}\right]\text{exp}\left[-\frac{H(\mathcal{J})}{T}\right] \\ &= \sum_{\mathcal{J}} W(\mathcal{J} \mid \mathcal{I})\text{exp}\left[-\frac{H(\mathcal{I})}{T}\right] \\ &= \text{exp}\left[-\frac{H(\mathcal{I})}{T}\right]\sum_{\mathcal{J}} W(\mathcal{J} \mid \mathcal{I}) = \text{exp}\left[-\frac{H(\mathcal{I})}{T}\right]\end{aligned}$$

which shows that the Boltzmann distribution is the eigenvector of $\mathbf{W}$ associated with the eigenvalue $\lambda = 1$ and therefore that it is the steady distribution. The detailed balance principle suggests a stochastic dynamics, known as the Metropolis dynamics, which is embedded in the following algorithm :

Begin :
1) Let $\mathcal{J}(\nu)$ be the state at time ν. Choose a state $\mathcal{I}$
2) If $H(\mathcal{I}) < H(\mathcal{J})$ then take it with probability 1 : $\mathcal{J}(\nu + 1) = \mathcal{I}$
3) If $H(\mathcal{I}) > H(\mathcal{J})$ then take it with probability $\text{exp}\left[-\frac{H(\mathcal{I}) - H(\mathcal{J})}{T}\right]$
4) Iterate in (1)
End.

One observes that

$$\frac{W(\mathcal{J} \mid \mathcal{I})}{W(\mathcal{I} \mid \mathcal{J})} = \frac{\exp\left[\frac{H(\mathcal{I}) - H(\mathcal{J})}{T}\right]}{1}$$

and therefore that the Metropolis algorithm respects the detailed balance principle.

In reality the models of neural networks imply more than the conditions (16). They also involve a dynamics which consists in choosing an unit at random and in updating its state according to (see eq. 16) :

$$\sigma_i(\nu+1) = \mathrm{Sgn}\left(\sum_{j=0}^{N} J_{ij}\sigma_j(\nu)\right) \tag{22}$$

In the limit $T \to 0$ the Metropolis algorithm reduces to :

$$\sigma_i(\nu+1) = \mathrm{Sgn}\left[\sigma_i(\nu)(H(\mathcal{I}) - H(\mathcal{J}))\right] \tag{23}$$

In general the neuronal dynamics (22) cannot be put in the form (23) which means that the neuronal dynamics does not obey the detailed balance principle, therefore that the asymptotic distribution is not that of Boltzmann and finally that the dynamics does not tend to make the free energy minimal.

Let us study the conditions which would make the neuronal dynamics obey the detailed balance equation. It is first necessary to use the cost function (20). As one single neural state is updated at every time step the states $\mathcal{I}$ and $\mathcal{J}$ differ only by the state $\sigma_i \equiv \sigma_i(\mathcal{J}) = -\sigma_i(\mathcal{I})$ and

$$H(\mathcal{I}) - H(\mathcal{J}) = -\sum_i (\gamma_i(\mathcal{I}) - \gamma_i(\mathcal{J})) = 2\sigma_i \sum_j (J_{ij} + J_{ji})\sigma_j \tag{24}$$

It compels the state σ_i to align along the direction of $\sum(J_{ij} + J_{ji})\sigma_j$ with probability 1 :

$$\sigma_i(\nu+1) = \mathrm{Sgn}\left(\sum_{j=0}^{N}(J_{ij} + J_{ji})\sigma_j(\nu)\right) = \mathrm{Sgn}\left(\sum_j J_{ij}^S \sigma_j(\nu)\right) \tag{25}$$

Where

$$\mathbf{J}^S = \frac{1}{2}\left(\mathbf{J} + \mathbf{J}^T\right)$$

is the symmetrical part of the matrix of connexions. To make the neuronal dynamics eq. (22) identical to the "thermodynamical dynamics" given in eq. (23) it is necessary that $\mathbf{J} = \mathbf{J}^S$, that is to say the connections must be symmetrical.

$$J_{ij} = J_{ji} \tag{26}$$

Then the cost function is called an Hamiltonian. If one strives to carry out the same reasoning on the cost function given by eq. (17) one finds :

$$\sigma_i(\nu + 1) = \mathrm{Sgn}\left[\sigma_i(\nu)\left(\sum_j (\theta(-\gamma_j(\mathcal{I})) - \theta(-\gamma_j(\mathcal{J})))\right)\right] \tag{27}$$

and there is no way of reducing this dynamics to the usual dynamics of neural networks.

In reality what the dynamics of eq. (22) strives to do is minimizing a "local energy"

$$H_i(\mathcal{I}) = -\gamma_i(\mathcal{I})$$

whereas the initial problem was to minimize a global energy $H(\mathcal{I})$ that is to say a function which takes all units into account at a time. The problem is that the "thermodynamical dynamics" given by eq. (25), which only takes the symmetrical part of neuronal interactions into account, introduces metastable states even in systems such as feed-forward networks which is avoided when the plain neuronal dynamics is used.

To summarize this section

- either the interactions are symmetrical. The cost function takes the form of a quadratic Hamiltonian :

$$H(\mathcal{I}) = -\sum_{\langle ij\rangle} J_{\langle ij\rangle}\sigma_i\sigma_j \tag{28}$$

where the summation is over all *pairs* $\langle ij\rangle$ of connexions. The neuronal and thermodynamical dynamics are then each other identical and the results and theorems of statistical mechanics are applicable to the asymptotic distribution brought about by the neuronal dynamics.

- or they are not. There still exists an asymptotic distribution associated with the neuronal dynamics but we have no general results regarding this distribution.

It is tempting then, to widen the class of neural networks to systems whose dynamics is *defined* by eq. (23) :

$$\sigma_i(\nu + 1) = \mathrm{Sgn}\left[\sigma_i(\nu)\left(H(\mathcal{I}) - H(\mathcal{J})\right)\right] \tag{23}$$

with :

$$H(\mathcal{I}) = \sum_{i=1}^{N} \mathcal{S}(-\gamma_i(\mathcal{I})) \tag{29}$$

$\mathcal{S}(x)$ is some monotonous increasing function. For example taking $\mathcal{S}(x) = x$ yields the cost given in eq.(20) and $\mathcal{S}(x) = \theta(x)$ leads to eq. (18). Noise may be introduced in eq. (23) by adding a random contribution η to the argument of the Sgn function :

$$\sigma_i(\nu + 1) = \mathrm{Sgn}\,[\sigma_i(\nu)\,(H(\mathcal{I}) - H(\mathcal{J})) + \eta] \tag{23'}$$

If T is the width of the distribution of η, the stochastic dynamics builds a steady distribution given by :

$$\rho(\mathcal{I}) = \frac{1}{Z}\exp\left(-\frac{H(\mathcal{I})}{T}\right)$$

and the tools of statistical mechanics become available for all sorts of networks. When the cost is given the form (20), a cost based on stabilization parameters, equation (25) shows that the steady distribution of dynamics (23') only retains the symmetrical part of the cost function, which amounts to saying that one can add any fully asymmetrical contribution to the set of interactions without modifying the asymptotic distribution. In other word choosing a "thermodynamical dynamics" leads to the lost of a part of information that is embedded in the set of interactions. What is actually lost if an error based cost function (18) is used instead of function (20) is not clear.

3. Learning and Generalization: General Statistical Approaches

3.1. A STATISTICAL MECHANICS APPROACH OF LEARNING

As one views the networks as machines aiming at solving problems it is necessary to distinguish *input units* whose states σ_j, $j = 1, \cdots, \mathcal{N}_I$, code for the data and *output units* σ_i, $i = 1, \cdots, \mathcal{N}_O$, which code for the answers. The set of input and ouput units makes the set of ***visible units***. The units $\sigma_h, h = 1, \cdots, \mathcal{N}_H$, which do not belong to the set of visible units are called ***hidden units***. The networks with $\mathcal{N}_H = 0$, that is to say the networks which do not comprise any hidden unit, are called ***visible networks*** (Fig. 2).

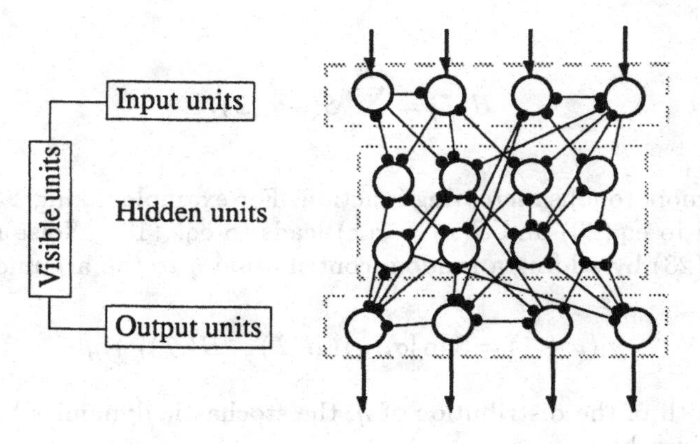

Figure 2. Defining the types of units.

The system is embedded into an environment which determines the input patterns that it is likely to experience. Some are more frequent than others, some are never generated by the environment. Moreover the selective pressure determines which response, which output pattern, is best adapted to a given input pattern. All in all the environment determines a probability distribution $\rho^E(I^v)$ of visible units. Most approaches of learning and generalization consider the following particular distribution :

$$\begin{array}{ll} \rho^E(I^v) = 1/P & \text{if } I^v \in \mathcal{E} \\ \rho^E(I^v) = 0 & \text{if } I^v \notin \mathcal{E} \end{array}$$

where $\mathcal{E}$ is called a *training set*. It is made of P visible patterns I^μ, $\mu = 1, \cdots, P$, with

$$I^\mu = I^{\mu,in} \otimes I^{\mu,out} \quad , \; I^{\mu,in} = \{\xi_j^\mu\} \;\; ; \; j =, \cdots, \mathcal{N}_I \;\; , \; I^{\mu,out} = \{\xi_i^\mu\} \;\; ; \; i = 1, \cdots, \mathcal{N}_O$$

The two main capabilities one usually wants an automata network to display are *learning* and *generalization.*

Learning is a process which strives to make the system behaves as expected for a subset of all possible visible patterns. Generalization is the property of a system, that has been trained with the subset, to behave for any input pattern according to the rules that environment uses to generate the probability distribution $\rho^E(I^v)$. Let us consider the problem of learning first. Depending on whether the environmental probability distribution is given its most general form or it is given in the form of a training set, two learning mechanisms have been put forward.

a) Using Kullback distances.

The idea is that the probability for a ***visible pattern*** $I^{\mu,v}$ to be spontaneously generated by the network in the ***unconstrained network*** must be equal to the probability that it occurs in the environment. An unconstrained network is a network whose visible units are not influenced by external stimuli. In other words the algorithm strives to match the internal representation of the neural network, ***observed on visible units***, with the states that environment generates on the visible units.

Let $\tilde{\rho}(I) \equiv \tilde{\rho}(I^v \otimes I^h)$ be the asymptotic distribution of the unconstrained network. The distribution probability $\rho^{un}(I^v)$ of the visible units in the unconstrained network is obtained by summing out the states of hidden units :

$$\rho^{un}(I^v) = \sum_{\{I^h\}} \rho(I^v \otimes I^h) \tag{30}$$

The suffix "un" is for "unconstrained networks". According to eq. (14) one defines a cost as the relative entropy F of the unconstrained distribution with respect to that of the stimuli. The learning algorithm will strive to make the relative entropy vanish. F is defined by :

$$F = T \sum_{\{I^v\}} \rho^E(I^v) \operatorname{Log}\left(\frac{\rho^E(I^v)}{\rho^{un}(I^v)}\right) \tag{31}$$

We have seen that F is positive for all probability distributions except when the two distributions ρ^{un} and ρ^E are identical :

$$\rho^{un}(I^v) \equiv \rho^E(I^v)$$

F is a function of the parameters $\{J\}$ of the system, $F = F(\mathbf{J})$. Learning then consists in modifying the parameters so as to make F eventually vanish. This may be achieved by using gradient algorithms in the space of parameters $\{J\}$. The learning dynamics therefore is written as :

$$\mathbf{J}(\nu+1) = \mathbf{J}(\nu) - \varepsilon \frac{\partial}{\partial \mathbf{J}} \cdot F(\mathbf{J}) \quad ; \; \varepsilon > 0 \tag{32}$$

This learning algorithm makes the relative entropy decrease since

$$dF = \left(\frac{\partial F}{\partial \mathbf{J}}\right)^T . d\mathbf{J} = \varepsilon \left|\frac{\partial F}{\partial \mathbf{J}}\right|^2 \leq 0$$

This is the principle of the "Boltzmann machine", a learning dynamics that has been imagined by Hinton and Sejnowski [8]. The algorithm is developped in section 3.2.

b) Introducing cost functions

When the environmental distribution is given in the form of a training set it is natural to directly compare the states that are observed on the output units that are triggered by the input patterns $\{I^{\mu,in}\}$, $\mu = 1, \cdots, P$ of the training set $\mathcal{E}$, with the desired output patterns $\{I^{\mu,out}\}$ and define a cost function accordingly. Two definitions are possible :

1) Supervised learning

The performance of a network for pattern $\mathcal{I}^{\mu}$ may be measured by the *number of successes (the profit)* observed on output units. This number $H^P(\mathcal{I}^{\mu})$ is given by :

$$H^P(\mathcal{I}^{\mu}) = \sum_{i=1}^{\mathcal{N}_O} \theta\left[\gamma_i^{\mu}\right] \quad \text{with} \quad \gamma_i^{\mu} = \xi_i^{\mu} h_i = \xi_i^{\mu} \sum_j J_{ij} \sigma_j(\mathcal{I}^{\mu}) \tag{33}$$

2) Associative learning

The performance is measured by the stability parameter :

$$H^P(\mathcal{I}^{\mu}) = \sum_{i=1}^{\mathcal{N}_O} \gamma_i^{\mu} \tag{34}$$

Learning is then a process where one strives *to make the worst case as good as possible.* If one succeeds making the system give convenient responses even for this worst case then one is certain that the system will work for all other cases. Learning is therefore a two-step procedure, close to classical mini-max optimization processes.

The first step consists in computing the average minimal profit. It is given by the free energy $F(\{J\})$ of systems whose sets of parameters $\{J\}$ are given. The free energy is computed by using one of the profit functions that have been defined above :

$$F(\{J\}) = -T\text{Log}(Z(\{J\}) \quad ;$$
$$Z(\{J\}) = \sum_{\{\sigma(\mathcal{I}^{\mu})\}} \exp\left[-\frac{H^P(\mathcal{I}^{\mu}) + \lambda H(\mathcal{I}^{\mu})}{T}\right] \quad ; \lambda > 0 \tag{35}$$

The second term in the exponent is given by eqs. (17) or (20). Its role is to compel the system to follow a thermodynamical dynamics. It is reminded that this dynamics is identical to the neuronal dynamics if the cost H is that given by eq. (20) with symmetrical interactions. $F(\{J\})$ is the free energy of a system with parameters $\{J\}$. It is written as :

$$F(\{J\}) = U(\{J\}) - TS(\{J\})$$

$U(\{J\})$, the internal energy of a system with parameters $\{J\}$, is the average profit. It is given by :

$$U(T,\{J\}) = \frac{d(\text{Log}(Z(\{J\})}{d(-1/T)} = \frac{d((-1/T)F(\{J\})}{d(-1/T)}$$

Letting T go to 0 gives $U(0,\{J\})$, the minimal profit that the particular machine defined by the parameters $\{J\}$, is able to achieve. The entropy :

$$S(0,\{J\}) = (-1/T)\,[F(0,\{J\}) - U(0,\{J\})]$$

is related to the degeneracy of the state corresponding to minimal profit. It is given by the logarithm of the number $\Gamma(\{J\})$ of such ground states $\{\sigma^{\mu}\}$.

$$S(0,\{J\}) = \text{Log}(\Gamma(\{J\}) \tag{36}$$

Here three strategies are possible according to the purpose of the theory.
1) If one wants to devise a learning dynamics, that is to say to find a set of parameters $\{\mathbf{J}\}$ which maximizes the profit one appeals to a gradient dynamics in the phase space of parameters : Finding the minimal free cost is finding the set of parameters $\{J\}$ which makes the profit $F(\{J\})$ as large as possible. The learning dynamics therefore is written as :

$$\mathbf{J}(\nu+1) = \mathbf{J}(\nu) + \varepsilon\frac{\partial}{\partial \mathbf{J}}\cdot F(\mathbf{J}) \quad ; \quad \varepsilon > 0 \tag{37}$$

This learning algorithm makes the free profit increase since

$$dF = \left(\frac{\partial F}{\partial \mathbf{J}}\right)^{T}.d\mathbf{J} = \varepsilon\left|\frac{\partial F}{\partial \mathbf{J}}\right|^{2} \geq 0$$

An application to the perceptron architecture is given in section 4.1.
2) If the theory aims at finding some general property as regards the “learnability” or the “generalizability” that is attached to a particular training set one carries out thermal averages in the phase space of parameters. As we have already stated, finding sets of parameters $\{J\}$ which make the profit as large as possible is equivalent to finding the parameters which minimize a cost defined by :

$$H'(\{J\}) = -F(\{J\})$$

The free cost is then given by :

$$F = -T'\text{Log}(Z) \quad \text{with} \quad Z = \sum_{\{J\}} \exp\left[+\frac{F(\{J\}}{T'}\right] \tag{38}$$

T' is the temperature of the thermal bath associated with dynamical variables $\{\mathbf{J}\}$ whereas T is the temperature of the thermal bath of dynamical variables $\{\sigma_i\}$. There is no reason for the two temperatures to be identical. In actual fact it is wise to take $T \gg T'$ to make sure that all patterns of the training set are evenly learned by the system.

3) Finally one could be interested by general properties, properties not depending on the particular set of patterns that has been chosen for training. An average over realization has then to be carried out. The relevant quantity is :

$$F = -T'\overline{\mathrm{Log}(Z)}$$

The computation of the average generally appeals to the *replica technique*. The performance of an ensemble of networks is given by :

$$U(T') = \frac{d((-1/T')F)}{d(-1/T')}$$

Taking the limit $T' \to 0$, yields the maximal profit (or, equivalently the minimal cost) U that is globally minimal. The entropy :

$$S(T'=0) = \mathrm{Log}(\Gamma) \tag{39}$$

gives the number of sets of parameters $\{J\}$ which achieve this maximal profit. The problem of learning has a solution if the number of errors that is associated with maximal profit is vanishes . This amounts to saying that there exists at least a set of parameters $\{J\}$ such that, for each pattern I^μ, there exists a hidden state $I^{\mu,hid} = \{\sigma_h^\mu\}$ which makes all errors vanish. The set of hidden states$\{I^{\mu,hid}\}$ is called the *internal representation* of the network. The entropy (38) gives the number of solutions. Γ is called the *volume of solutions.* It is interesting to note that the present approach gives approximate solutions even though the internal energy (the minimal number of errors) does not vanish, that is to say even though no exact solution exists.

3.2. THE BOLTZMANN MACHINE

The *"Boltzmann machine"* algorithm of Hinton and Sejnowski [8] is an attempt to build relevant internal representations in neural networks whose dynamics is driven by a quadratic symmetrical cost function eq. (28).

We assume that the stationary distribution of states $\rho(I,\infty) = \rho(I) \equiv \rho(I^v \otimes I^h)$ for the *unconstrained network*, that is to say for the network whose visible units are disconnected from the environment, is given by a Boltzmann distribution (whence the name) :

$$\rho(I) = \frac{1}{Z}\exp - \frac{1}{T}H(I) \quad \text{with} \quad Z = \sum_{\{I\}} \exp - \frac{1}{T}H(I) \tag{40}$$

with :

$$H(I) = -\sum_{\langle lm \rangle} J_{\langle lm \rangle}\sigma_l(I)\sigma_m(I)$$

The distribution probability $\rho^{un}(I^v)$ of the visible units in the unconstrained network is given by :

$$\rho^{un}(I^v) = \sum_{\{I^h\}} \rho(I^v \otimes I^h)$$

and the learning dynamics is built so as to make the Kullback distance between $\rho^E(I^v)$ and $\rho^{un}(I^v)$ eventually vanish (see eqs. 30 and 31). The derivative of the relative entropy with respect to the synaptic efficacies J_{ij} is given by :

$$\frac{\partial F}{\partial J_{ij}} = -T\sum_{\{I^v\}} \frac{\rho^E(I^v)}{\rho^{un}(I^v)} \frac{\partial \rho^{un}(I^v)}{\partial J_{ij}}$$

since the distribution $\rho^E(I^v)$ is fixed and therefore does not depend on the set of parameters J_{ij}. The asymptotic distribution of the unconstrained network is a Maxwell-Boltzmann distribution and, therefore :

$$\frac{\partial F}{\partial J_{ij}} = -T\sum_{\{I^v\}} \frac{\rho^E(I^v)}{\rho^{un}(I^v)} \frac{\partial}{\partial J_{ij}} \left(\frac{1}{Z} \sum_{\{I^h\}} \exp\left(\frac{1}{T} \sum_{\langle lm \rangle} J_{\langle lm \rangle}\sigma_l\sigma_m \right) \right)$$

$$\frac{\partial F}{\partial J_{ij}} = -T\sum_{\{I^v\}} \frac{\rho^E(I^v)}{\rho^{un}(I^v)} \frac{1}{T}\frac{1}{Z} \left(\sum_{\{I^h\}} \sigma_i\sigma_j \exp\left(\frac{1}{T} \sum_{\langle lm \rangle} J_{\langle lm \rangle}\sigma_l\sigma_m \right) \right) + \cdots$$

$$+ \sum_{\{I^v\}} \frac{\rho^E(I^v)}{\rho^{un}(I^v)} \frac{1}{T}\frac{1}{Z^2} \left(\sum_{\{I^h\}} \exp \frac{1}{T} \sum_{\langle lm \rangle} J_{\langle lm \rangle}\sigma_l\sigma_m \right) \times \cdots$$

$$\times \left(\sum_{\{I \equiv I^v \otimes I^h\}} \sigma_i\sigma_j \exp \frac{1}{T} \sum_{\langle lm \rangle} J_{\langle lm \rangle}\sigma_l\sigma_m \right)$$

We use the following equalities :

$$\begin{aligned}\frac{1}{Z}\sum_{\{I^h\}}\sigma_i\sigma_j\exp\left(\frac{1}{T}\sum_{\langle lm\rangle}J_{\langle lm\rangle}\sigma_l\sigma_m\right) &= \sum_{\{I^h\}}\sigma_i\sigma_j\rho\left(I^v\otimes I^h\right)\\ &= \sum_{\{I^h\}}\sigma_i\sigma_j\rho\left(I^h\mid I^v\right)\rho^{un}\left(I^v\right)\\ &= \langle\sigma_i\sigma_j\rangle^{cons}\rho^{un}\left(I^v\right)\end{aligned}$$

where the suffix "*cons*" means that the average has to be carried out with visible units clamped on the various states of the training set, and :

$$\frac{1}{Z}\sum_{\{I^h\}}\exp\left(\frac{1}{T}\sum_{\langle lm\rangle}J_{\langle lm\rangle}\sigma_l\sigma_m)\right) = \sum_{\{I^h\}}\rho\left(I^v\otimes I^h\right) = \rho^{un}\left(I^v\right)$$

to obtain :

$$\frac{\partial F}{\partial J_{ij}} = -T\sum_{\{I^v\}}\rho^E\left(I^v\right)\frac{1}{T}\langle\sigma_i\sigma_j\rangle^{cons} + \sum_{\{I^v\}}\rho^E(I^v)\frac{1}{T}\langle\sigma_i\sigma_j\rangle^{un}$$

Finally, since

$$\sum_{\{I^v\}}\rho^E\left(I^v\right) = 1$$

we find :

$$\frac{\partial F}{\partial J_{ij}} = \left(\langle\sigma_i\sigma_j\rangle^{un} - \overline{\langle\sigma_i\sigma_j\rangle^{cons}}\right) \tag{41}$$

where

$$\overline{\langle\sigma_i\sigma_j\rangle^{cons}} = \sum_{\{I^v\}}\rho^E(I^v)\langle\sigma_i\sigma_j\rangle^{cons}$$

and the learning rule is :

$$\Delta J_{ij} = -\varepsilon\frac{\partial F}{\partial J_{ij}} = \varepsilon\left(\langle\sigma_i\sigma_j\rangle^{un} - \overline{\langle\sigma_i\sigma_j\rangle^{cons}}\right) \quad \text{with} \quad \varepsilon > 0 \tag{42}$$

In actual fact the patterns are learned after one another : The learning dynamics is made serial :

$$\Delta J_{ij}(\mu) = \varepsilon\left(\langle\sigma_i^\mu\sigma_j^\mu\rangle^{un} - \langle\sigma_i^\mu\sigma_j^\mu\rangle^{cons}\right) \quad \text{with} \quad \varepsilon > 0$$

To compute both correlation functions the initial state of the system is set to :

$$I(\nu = 0) = I^{\mu,v} \otimes I^h$$

where I^h is a random hidden state. The second correlation is computed by clamping the visible state to I^μ and letting the hidden units relax, while the first is computed by leaving all neurons free to relax.

3.3. A FEW COMMENTS ON THE MEANING OF GENERALIZATION

Generalization is the property that makes the automata networks useful at solving difficult problems. Let us assume that learning is defined by some training set $\mathcal{E}$ and that the patterns I^μ of the training set are generated by using definite rules, for example the output state $I^{\mu,out}$ is the result of a Boolean function applied to input pattern $I^{\mu,in}$. The training set is a subset of all patterns that may be generated that way. One says that a system displays generalization capabilities if it yields the convenient response I^{out} to an input I^{in} not belonging to the training set. If the case arises the learning stage has trapped the unknown rules, (it materializes the Boolean function), in the set of parameters $\{J\}$ it has built.

So exposed the problem of generalization is meaningless. Actually let us assume that the training set is made of P patterns we choose at will and that learning may be successfully achieved for all patterns. The generalization capabilities of the network after it has been trained with the $P-1$ first patterns are certainly very poor since we can choose a response $I^{\mu=P,out}$ to input $I^{\mu=P,in}$ that is quite different from the observed output I^{out} triggered by the input $I^{\mu=P,in}$. Therefore for a system to show generalization capabilities it is necessary that the examples of the training set cannot be chosen at will and that the system is not able to learn any set of examples. The first condition is met by assuming that the system (the rules) which generates the examples is unable to implement any Boolean function but only a subset of the set of possible Boolean functions. This means that among the population of possible patterns some patterns are more likely than others. This is materialized in the environment dependent distribution $\rho^E(I^v)$ we have introduced above. $\rho^E(I^v)$ may also be considered as the probability that the pattern which is to be learned next is I^v.

The second condition that the system cannot learn any set of patterns is also compulsory. It is satisfied by assuming that the architecture of the system cannot be chosen at will but that we have to pick it up among a set of possible architectures. One therefore introduces another probability, $\rho^0(\{J\})$, which is the *a priori* distribution of network parameters.

3.4. A THEORY OF GENERALIZATION USING THE ANNEALED APPROXIMATION

We now explain a general theory of generalization that has been put forward by Levin, Tishby and Solla [13]. Let $F^{(P)}(\{J\})$ be the average free cost (the negative of the free profit) of a network whose parameters are $\{J\}$ for a set of P training patterns. According to eq.(38) and taking the *a priori* probability of connexions $\rho^0(\{J\})$ into account, the probability for a network that its set of parameters is $\{J\}$ is given by :

$$\rho^{(P)}(\{J\}) = \frac{1}{Z^{(P)}}\rho^0(\{J\})\exp\left[-\frac{1}{T'}F^{(P)}(\{J\})\right] \quad (43)$$

with

$$Z^{(P)} = \int d\{J\}\ \rho^0(\{J\})\exp\left[-\frac{1}{T'}F^{(P)}(\{J\})\right]$$

F is an extensive quantity : If a new pattern is added to the training set the average free cost for the $P+1$ patterns is sum of the free cost for the P first patterns and the cost brought about by the new pattern I^v :

$$F^{(P+1)}(\{J\}) = F^P(\{J\}) + F(I^v, \{J\})$$

The probability

$$\rho(I^v, \{J\}) = \frac{1}{z}\exp\left[-\frac{F(I^v, \{J\})}{T'}\right] \quad (44)$$

with

$$z = \int d\{J\}\ \exp\left[-\frac{F(I^v, \{J\})}{T'}\right]$$

may be considered either as the probability for a network whose training set is only comprised of the new pattern to have $\{J\}$ as its set of parameter, or as the probability for the pattern to be realized by a network with parameter set $\{J\}$. The latter point of view is adopted in the following reasoning.

The additivity of costs allows the probability $\rho^{(P)}(\{J\})$ to be written as :

$$\rho^{(P)}(\{J\}) = \frac{1}{Z^{(P)}}\rho^0(\{J\})\prod_{\mu=1}^{P}\exp\left[-\frac{F(I^\mu, \{J\})}{T'}\right] \quad (45)$$

with :

$$Z^{(P)} = \int d\{J\}\ \rho^0(\{J\})\prod_{\mu=1}^{P}\exp\left[-\frac{F(I^\mu, \{J\})}{T'}\right]$$

The probability $\rho^{(P)}(I^v)$ that a state I^v not belonging to the training set $\mathcal{E}$, is produced by a network that has been trained by P patterns (that is the generalization capability of the network) is :

$$\rho^{(P)}(I^v) = \int d\{J\} \; \rho^{(P)}(\{J\}) \rho(I^v, \{J\})$$

Using the eqs. (43) and (44), the sum is transformed into :

$$\begin{aligned} \rho^{(P)}(I^v) &= \frac{1}{z(I^v)Z^{(P)}} \int d\{J\} \; \rho^0(\{J\}) \exp\left[-\frac{F^{(P)}(\{J\}) + F(I^v, \{J\})}{T'}\right] \\ &= \frac{Z^{(P+1)}}{z(I^v)Z^{(P)}} \end{aligned}$$

This expression has to be averaged over all training sets of P patterns. The average over realizations is in general a task which is difficult to carry out. A simple approximation, called the "annealed approximation", consists in assuming that the averaging process can be carried out on partition functions Z rather than on free energies $F = -T\mathrm{Log}(Z)$. Moreover the average of $Z^{(P)}$, $Z^{(P+1)}$ and z are one another decoupled leading to :

$$\overline{\rho^{(P)}} = \frac{\overline{Z^{(P+1)}}}{\overline{z}\,\overline{Z^{(P)}}} \tag{46}$$

where the partial partition function z which depends on I^v has been replaced by its average over all possible I^v.

On the other hand it is reminded that the set of visible patterns is distributed along $\rho^E(I^v)$. For a network whose set of parameters is $\{J\}$, an average generalization capacity $g(\{J\})$ is defined as :

$$g(\{J\}) = \int d\{I^v\} \; \rho^E(I^v) \rho(I^v, \{J\})$$

and a generalization density $\rho^{(P)}(g)$ is introduced :

$$\rho^{(P)}(g) = \int d\{J\} \; \rho^{(P)}(\{J\}) \delta(g(\{J\}) - g)$$

The generalization density is to be averaged over the realizations of visible patterns leading to :

$$\begin{aligned} \overline{\rho^{(P)}(g)} = \frac{1}{\overline{Z^{(P)}}} \int \prod_{\mu=1}^{P} d(I^{\mu,v}) \; \rho^E(I^{\mu,v}) \cdots \\ \cdots \int d\{J\} \; \rho^0(\{J\}) \exp\left[-\frac{F(I^{\mu,v}, \{J\})}{T'}\right] \delta(g(\{J\}) - g) \end{aligned}$$

where it has been assumed, once again, that the averages can be decoupled. Introducing the equations (44) and (45) in the last expression yields :

$$\begin{aligned} \overline{\rho^{(P)}(g)} &= \frac{1}{\overline{Z^{(P)}}} \overline{z}^P \int d\{J\}\ \rho^0(\{J\})(g(\{J\}))^P\ \delta(g(\{J\}) - g) \\ &= \frac{1}{\overline{Z^{(P)}}} \overline{z}^P g^P \rho^0(g) \end{aligned} \tag{47}$$

where the distribution

$$\rho^0(g) = \int d\{J\}\ \rho^0(\{J\})\delta(g(\{J\}) - g)$$

is the *a priori* generalization density. Its contains all information about the architecture through the distribution $\rho^0(\{J\})$ and about the function the system has to implement through the definition of $g(\{J\})$. The average generalization capacity is given by :

$$\begin{aligned} \overline{\rho^{(P)}} &= \int_0^1 dg\ g\overline{\rho^{(P)}(g)} \\ &= \frac{\overline{z}^P}{\overline{Z^{(P)}}} \int_0^1 dg\ \rho^0(g) g^{P+1} \end{aligned} \tag{48}$$

Making eq. (48) identical to eq. (46) yields :

$$\overline{Z^{(P+1)}} = \overline{z}^P \int_0^1 dg\ \rho^0(g) g^{P+1}$$

and, finally :

$$\overline{\rho^{(P)}} = \frac{\int_0^1 dg\ \rho^0(g) g^{P+1}}{\int_0^1 dg\ \rho^0(g) g^P} \tag{49}$$

The conclusion is that, in the annealing approximation, the average generalization capability of the network is simply given by the ratio of successive moments of the *a priori* generalization density. For $P \to \infty$ the generalization capability is fully determined by the behaviour of $\rho^0(g)$ in the vicinity of $g \simeq 1$. If $\rho^0(g) \simeq (1-g)^{Nd}$ with $d \geq 0$ then :

$$\overline{\rho^{(P)}} \simeq 1 - \frac{Nd+1}{P} \simeq 1 - \frac{d}{\alpha} \quad \text{with} \quad \alpha = P/N \tag{50}$$

and the generalization error vanishes as α^{-1}.

4. Visible Networks

4.1. PERCEPTRONS: COMPUTING THE ZERO NOISE ENTROPY

In visible networks one exactly knows what the states of minimal cost should be, at least at zero noise level : They are fully determined by the states I^μ, $\mu = 1, \cdots, P$, of the training set $\mathcal{E}$. In more general networks one *a priori* ignores how to choose the state $I^{\mu,h}$ of hidden units to be associated with a certain pattern I^μ. This is the problem of *credit assignment*. This chapter deals with visible networks exclusively.

We consider a particular unit i of a visible network. If

$$\mathrm{Sgn}\left(\sum_{j=0}^{N} J_{ij}\xi_j^\mu\right) = \sigma_i(I^\mu) \neq \xi_i^\mu$$

one can say that one error is associated with the subnetwork made of N input units $j = 1, \cdots, N$ projecting onto a unique output unit i, for pattern I^μ. These subnetworks are called *perceptrons*. A visible network may be considered as a set of N intertwined perceptrons i, $i = 1, \cdots, N$ (Fig. 3).

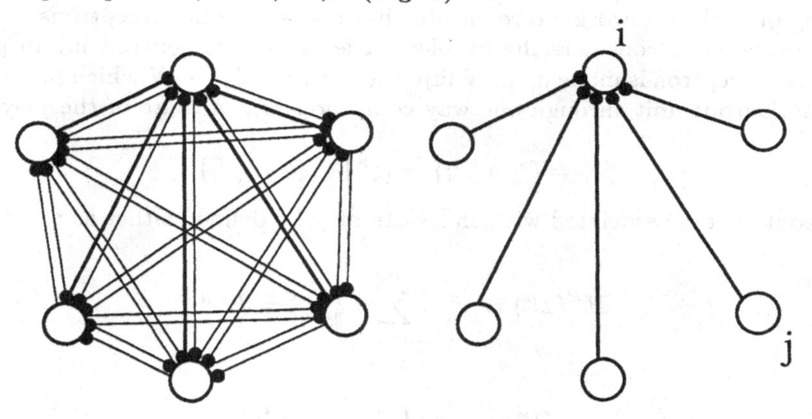

Figure 3. Visible networks: A set of perceptrons.

Strictly speaking, a perceptron is feed-forward which is not the case if the unit i is self-connected. To escape this problem one assumes that $J_{ii} = 0$. The total number of errors for perceptron i is given by:

$$H_i = \sum_{\mu=1}^{P} \theta\left[-\gamma_i^\mu\right]$$

with :

$$\gamma_i^\mu = \xi_i^\mu h_i(I^\mu) = \sum_{j=0}^{N} J_{ij}\xi_i^\mu \xi_j^\mu$$

Let us assume that one finds a learning dynamics which produces optimal connections (minimal H_i). Since the plain neuronal dynamics strives precisely to minimize H_i, the states it generates will reproduce this minimization process, that is to say ξ^μ will be the ouput of I^μ which is what is desired.

On the other hand the overall number of errors is

$$H = \sum_i H_i$$

Since the J_{ij}'s (i given), which minimize each H_i are one another decoupled for different i, the set $\{J\}$ of parameters which makes H minimal is the union of sets of all weights found for each perceptron. Moreover, since the state dynamics which minimizes H is the symmetrized one given by eq. 24 one concludes that the symmetrized dynamics as well as the plain neuronal dynamics will find the same states of minimal energies, namely the memorized patterns I^μ and the problem of learning in visible networks to reduce to that of learning in perceptrons.

One therefore considers the problem of learning and generalizing in perceptrons. A perceptron is made up of N input units i, $i = 1, \cdots, N$ which project onto a unique output unit through one way connexions J_i. A state of the perceptron is :

$$\mathcal{I}^\mu = I^\mu \otimes \{J_i\} = \{\xi_i^\mu\} \otimes \xi^\mu \otimes \{J_i\}$$

The profit that is associated with this state may be defined either as :

$$H^P(\mathcal{I}^\mu) = \gamma^\mu = \sum_{i=1}^{N} J_i \xi_i^\mu \xi^\mu = \tilde{J}.\tilde{\tau}^\mu \tag{51}$$

with

$$(\tilde{J})_i = J_i \quad \text{and} \quad (\tilde{\tau}^\mu)_i = \xi_i^\mu \xi^\mu$$

or as :

$$H^P(\mathcal{I}^\mu) = \theta(\gamma^\mu) \tag{52}$$

for all patterns $\mathcal{I}^\mu$ belonging to the training set $\mathcal{E}$ and

$$H^P(\mathcal{I}) \to +\infty$$

for all other patterns.

According to the minimax learning procedure of paragraph 2 one first computes the partition function $Z(\tilde{J})$ by summing over all states :

$$Z(\tilde{J}) = \sum_{\mu=1}^{P} \exp[-\frac{\tilde{J}.\tilde{\tau}^{\mu}}{T}] \text{ or } Z(\tilde{J}) = \sum_{\mu=1}^{P} \exp[-\frac{\theta(\tilde{J}.\tilde{\tau}^{\mu})}{T}] \quad (53)$$

Then the free energy $F(\tilde{J})$ is given by :

$$F(\tilde{J}) = -T\text{Log}(Z(\tilde{J})) \quad (54)$$

The summation over the parameters $\tilde{J}$ is carried out :

$$Z = \sum_{\tilde{J}} \exp[+\frac{F(\tilde{J})}{T'}] \quad (55)$$

and the free energy of the ensemble of networks is given by :

$$F = -T'\text{Log}(Z) \quad (56)$$

In particular the average number of errors of the ensemble is given by :

$$\overline{H}(T') = \frac{d((-1/T')F)}{d(-1/T')} \quad (57)$$

and the entropy by :

$$S(T') = (-1/T')(F - \overline{H}) \quad (58)$$

The E. Gardner's derivation of the zero noise entropy

We focus our attention on the zero noise entropy $S(0)$ with a profit defined by eq. (52). For small but finite T's, according to the last of eqs. (53), $Z(\tilde{J})$ is close to the number of errors. Then, in the limit $T'/T \to 0$, Z tends to be the number Γ of vectors $\tilde{J}$ which determine the perceptrons with minimal number of errors. According to eq. (36) the entropy is then given by :

$$S = \text{Log}(\Gamma)$$

Let us assume that the minimal number of errors vanishes. Then Γ is the number of vectors $\tilde{J}$, the number of machines, which produce a correct answer $\sigma(I^\mu) \equiv \xi^\mu$ for all patterns I^μ. Γ is given by :

$$\Gamma = \sum_{\tilde{J}} \prod_{\mu=1}^{P} \theta\left[\tilde{J}.\tilde{\tau}^\mu\right] \tag{59}$$

For avoiding Γ to be become infinite it is necessary to put some constraints on the space that is spanned by the vectors $\tilde{J}$. The constraint :

$$\sum_i (J_i)^2 = N$$

amounts to using normalized vectors $\tilde{J}$. Renormalizing $\tilde{J}$ does not modify the number of errors. The quantity of interest is then the relative volume of solutions in this restricted phase space. It is given by :

$$\Gamma = \frac{\int \prod dJ_i [\prod_\mu \theta(\frac{1}{\sqrt{N}} \xi^\mu \sum_i J_i \xi_i^\mu)][\delta(\sum_i (J_i)^2 - N)]}{\int \prod dJ_i [\delta(\sum_i (J_i)^2 - N)]} \tag{60}$$

A complete calculation of S involves averaging over sets of training patterns, that is to say averaging over possible realizations of $\mathcal{E}$ (eq. 39) :

$$\overline{S} = \overline{\text{Log}(\Gamma)}$$

The derivation has been fully carried out by E. Gardner [4,5]. By using the identity :

$$\theta(a) = \int_0^\infty \frac{d\lambda}{2\pi} \int_{-\infty}^{+\infty} dx \, \exp(ix(\lambda - a))$$

E. Gardner observes that Γ is cast in a form fully ressembling an usual partition function. She therefore applies the tools of statistical mechanics to the calculation. In particular the sample averaging is carried out by appealing to the replica method. We shall not give the details of the derivation here which are rather lengthy. The result we arrive at is that the volume of solutions shrinks to zero when the relative number of patterns $\alpha = P/N$ is larger than a critical value α_c. For randomly generated patterns α_c is given by :

$$\alpha_c = \frac{1}{\int_0^{+\infty} \frac{dy}{\sqrt{2\pi}} y^2 \exp\left(-\frac{y^2}{2}\right)} = 2$$

In actual fact this result had been derived long beforehand by Cover who used a geometrical reasoning. However the derivation of E. Gardner is much more far reaching. Her technique can be applied to a great many problems. For example Gardner proved that the existence of correlations between the patterns makes the critical value α_c increase. Eventually α_c grows to infinity when the patterns tend to be fully aligned along one another, a not very intuitive result.

4.2. LEARNING DYNAMICS FOR PERCEPTRON ARCHITECTURES

The general principles we have introduced in chapter 3 are now applied to the perceptron architecture. Using the partition function $Z(\left\{\tilde{J}\right\})$ displayed in eq. (35) and the gradient algorithm eq. (37) one finds :

$$\frac{d\tilde{J}}{dt} = \varepsilon \frac{1}{Z(\{J\})} \sum_{\mu=1}^{P} \tilde{\tau}^{\mu} \exp[-\frac{\tilde{J}.\tilde{\tau}^{\mu}}{T}] \tag{61}$$

with

$$Z(\{J\}) = \sum_{\mu=1}^{P} \exp[-\frac{\tilde{J}.\tilde{\tau}^{\mu}}{T}]$$

This form of continuous learning has been proposed three years ago by one of us. It is instructive to see how it transforms in the limits of large and low temperatures T.

In the limit $T \to \infty$ all exponentials tend to 1. Then $Z \to P$ and the learning algorithm becomes :

$$\frac{d\tilde{J}}{dt} = \frac{\varepsilon}{P} \sum_{\mu=1}^{P} \tilde{\tau}^{\mu} \tag{62}$$

which is the Hebbian rule.

In the limit $T \to 0$ the sum of exponentials is dominated by the term which gives the lowest value of $\tilde{J}\tilde{\tau}^{\mu}$. Let $I^{\mu_{min}}$ be the corresponding pattern. The learning dynamics becomes :

$$\frac{d\tilde{J}}{dt} = \varepsilon \tilde{\tau}^{\mu_{min}} \tag{63}$$

which is the Minover algorithm of Krauth and Mezard [11]. One knows that the Hebbian rule is not optimal as far as pattern storage is concerned. Simulations have shown that the learning dynamicse (61) with temperature of the order of N is able to yield the theoretical maximal storage capacity $\alpha_c = 2$ indeed (for Hebbian rules $\alpha_c = 0.14$) (Fig. 4).

It may be shown that if $\alpha < \alpha_c = 2$ the length $\left|\tilde{J}\right|$ of $\tilde{J}$ grows logarithmically beyond any limit whereas $\tilde{J}$ tends towards a fixed point in the other case. This can be used as a mean of avoiding the overloading (the overcrowding catastrophe) of long term memorization. Nevertheless one could demand that the length of $\tilde{J}$ is fixed as in Gardner's calculation : $\left|\tilde{J}\right|^2 = N$. This can be carried out by merely introducing this norm in the definition of the partition function :

$$Z(\{J\}) = \sum_{\mu=1}^{P} \exp\left[-\frac{\tilde{J}\cdot\tilde{\tau}^\mu}{\left|\tilde{J}\right| T}\right].$$

and

$$\frac{d\tilde{J}}{dt} = \frac{\varepsilon}{Z(\{J\})\left|\tilde{J}\right|}\sum_{\mu=1}^{P}\exp\left[-\frac{\tilde{J}\cdot\tilde{\tau}^\mu}{\left|\tilde{J}\right| T}\right]\left[\tilde{\tau}^\mu - \frac{(\tilde{J}.\tilde{\tau}^\mu)\tilde{J}}{\left|\tilde{J}\right|^2}\right] \tag{64}$$

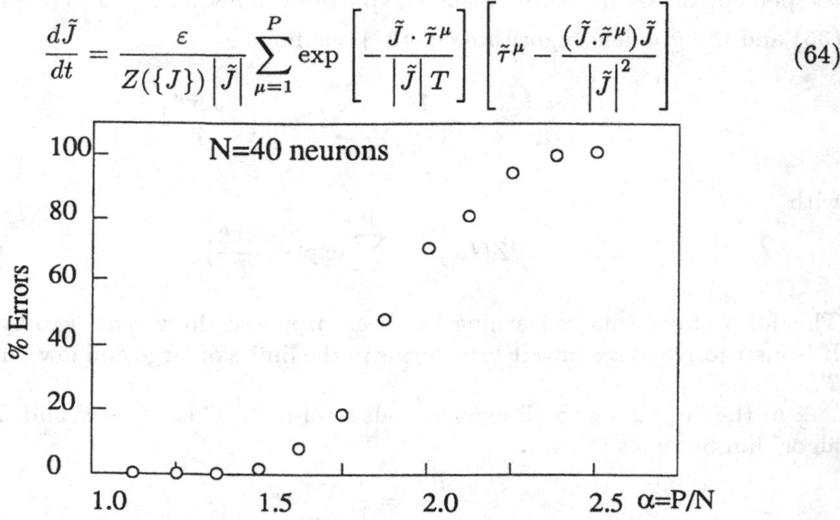

Figure 4. Learning using eq. 61 yields the maximum capacity $\alpha = 2$. The rounding of the transition is due to finite size effects.

One verifies that :

$$\tilde{J}.\frac{d\tilde{J}}{dt} = 0$$

which means that the rule (64) is norm-conserving. From eq. (64) the fixed points $\tilde{J}^*$ of the learning dynamics are solutions of the following equation :

$$\tilde{J}^* = \frac{\sum_{\mu=1}^{P}\tilde{\tau}^\mu \exp[-\frac{\tilde{J}^*.\tilde{\tau}^\mu}{T\sqrt{N}}]}{\sum_{\mu=1}^{P}\frac{\tilde{J}^*.\tilde{\tau}^\mu}{N}\exp[-\frac{\tilde{J}^*.\tilde{\tau}^\mu}{T\sqrt{N}}]}$$

4.3. GENERALIZATION IN HEBBIAN PERCEPTRONS

The following derivation has been put forward by F. Vallet [20].

One assumes that the set of possible patterns is generated by a teacher perceptron comprised of N input units. Let $\tilde{J}^T$ be the set of connections of the teacher perceptron. The possible patterns are then given by :

$$I^\mu = \tilde{\xi}^\mu \otimes \xi^\mu \quad , \quad \mu = 1, \cdots, 2^N$$

where $\tilde{\xi}^\mu$ is the input vector and ξ^μ, the output state, is determined by :

$$\xi^\mu = \mathrm{Sgn}\left[\tilde{J}^T.\tilde{\xi}^\mu\right]$$

A pupil perceptron strives to learn the Boolean function which is defined by the set of 2^N patterns I^μ. For that purpose it uses a random set $\mathcal{E}$ of P patterns I^μ to build its connections along the Hebbian rule. The set of connections, that is to say the result of learning, is :

$$\tilde{J}^P = \sum_{\mu=1}^{P} \tilde{\xi}^\mu \xi^\mu = \sum_{\mu=1}^{P} \tilde{\xi}^\mu \mathrm{Sgn}\left[\tilde{J}^T.\tilde{\xi}^\mu\right] \tag{65}$$

One chooses an input vector $\tilde{\xi}$ not belonging to $\mathcal{E}$, that is $\tilde{\xi} \notin \left\{\tilde{\xi}^\mu, \mu \le P\right\}$ and we compute the probability $\overline{\rho^{(\alpha)}}$, $\alpha = P/N$, that the pupil system yields the correct answer, that which is given by the teacher perceptron. If $\overline{\rho^{(\alpha)}} = 1$ generalization is perfect. $\overline{\rho^{(\alpha)}} = 1/2$ is the sign that the pupil perceptron is not apt at generalizing. The probability that the answer of the pupil is correct is the probability that :

$$\mathrm{Sgn}\left[\tilde{J}^P.\tilde{\xi}\right] \mathrm{Sgn}\left[\tilde{J}^T.\tilde{\xi}\right] > 0$$

which, with eq. (65), is the probability that :

$$\sum_{\mu=1}^{P=\alpha N} (\tilde{\xi}^\mu.\tilde{\xi})\mathrm{Sgn}\left[\tilde{J}^T.\tilde{\xi}^\mu\right] \mathrm{Sgn}\left[\tilde{J}^T.\tilde{\xi}\right] > 0 \tag{66}$$

One considers the distribution of the random vector $\tilde{\zeta}^\mu$:

$$\tilde{\zeta}^\mu = \tilde{\xi}^\mu \mathrm{Sgn}\left[\tilde{J}^T.\tilde{\xi}^\mu\right]$$

Its average over a random distribution of patterns is :

$$\overline{\tilde{\zeta}} = \sqrt{\frac{2}{\pi}} \frac{\tilde{J}^T}{\left|\tilde{J}^T\right|}$$

We write the equation (66) as :

$$\sum_{\mu=1}^{\alpha N} z^{\mu} > 0$$

That is to say we must compute the probability $\overline{\rho^{(P)}(\tilde{\xi})}$ that a sum of αN random variables with averages :

$$\overline{z} = \sqrt{\frac{2}{\pi}} \frac{(\tilde{J}^T.\tilde{\xi})\mathrm{Sgn}\left[\tilde{J}^T.\tilde{\xi}\right]}{\left|\tilde{J}^T\right|}$$

and mean square deviation :

$$\overline{(\Delta z)^2} = N$$

is positive. This is given by :

$$\begin{aligned} \overline{\rho^{(P)}(\tilde{\xi})} &= \frac{1}{\sqrt{2\pi NP}} \int_0^{\infty} dz \exp\left[-\frac{(z - P\overline{z})^2}{2NP}\right] \\ &= \frac{1}{N\sqrt{2\pi\alpha}} \int_{-\alpha N\overline{z}}^{\infty} dz \exp\left[-\frac{z^2}{2\alpha N^2}\right] \end{aligned}$$

With

$$u = \frac{z}{N\sqrt{2\alpha}}$$

and the definition:

$$\mathrm{erfc}(x) = \frac{2}{\sqrt{\pi}} \int_x^{\infty} du \exp\left[-u^2\right]$$

the probability is written as :

$$\overline{\rho^{(P)}(\tilde{\xi})} = \frac{1}{2}\mathrm{erfc}\left[-\sqrt{\frac{\alpha}{\pi}} \frac{(\tilde{J}^T.\tilde{\xi})\mathrm{Sgn}\left[\tilde{J}^T.\tilde{\xi}\right]}{\left|\tilde{J}^T\right|}\right] \qquad (67)$$

An average has now to be carried out over the possible input vectors $\tilde{\xi}$. One defines a new random variable v :

$$v = \frac{(\tilde{J}^T.\tilde{\xi})\mathrm{Sgn}\left[\tilde{J}^T.\tilde{\xi}\right]}{\left|\tilde{J}^T\right|}$$

Its distribution is given by :

$$\rho(v) = \theta(v)\sqrt{\frac{2}{\pi}}\exp\left[-\frac{v^2}{2}\right] \tag{68}$$

Using the equations (67) and (68) the generalization probability is finally given by :

$$\overline{\rho^{(\alpha)}} = \frac{1}{\sqrt{\pi}}\int_0^\infty dv\ \exp[-v^2]\mathrm{erfc}[-v\sqrt{\frac{2\alpha}{\pi}}] \tag{69}$$

Reintroducing the definiton of the erfc function in equation (69) leads to :

$$\overline{\rho^{(\alpha)}} = \frac{2}{\pi}\int_0^\infty dv\ \exp\left[-v^2\right]\int_{-v\sqrt{\frac{2\alpha}{\pi}}}^\infty du\ \exp\left[-u^2\right]$$

Using polar coordinates $r^2 = u^2+v^2$, $\theta = \arctan(v/u)$ makes the double integration simple. One finds :

$$\overline{\rho^{(\alpha)}} = 1 - \frac{1}{\pi}\arctan\left(\sqrt{\frac{\pi}{2\alpha}}\right) \tag{70}$$

One observes that the pupil perceptron generalizes very slowly. In actual fact the generalization error decreases as $P^{-1/2}$ for large values of α. This is to be compared with the P^{-1} behaviour that is predicted by the theory of Tishby [13] et al. (section 3.4). This is another indication that the Hebbian rule is not a very efficient learning algorithm. We shall see in the next section that efficient rules lead to much more powerful generalization capabilities.

4.4. GENERALIZATION IN BINARY PERCEPTRONS

A combinatorial approach

Binary perceptrons are perceptrons whose synaptic weights are constrained to values $J_i \in \{-1, +1\}$. The problem of generalization in binary perceptrons has been introduced and solved at zero temperature by Gardner and Derrida [5]. Here too, two perceptrons are considered, a teacher perceptron whose connection vector is $\tilde{J}^T$, ($J_i^T \in \{-1,+1\}$), and a pupil perceptron with connection vector $\tilde{J}^P$,($J_i^P \in \{-1,+1\}$). For the pupil perceptron to generalize perfectly it is necessary that $\tilde{J}^P \equiv \tilde{J}^T$ since a single error for the connections would bring about different responses for some inputs. One then defines a connection overlap m :

$$m = \frac{1}{N}\tilde{J}^T.\tilde{J}^P$$

Let φ be the angle between $\tilde{J}^T$ and $\tilde{J}^P$.

$$\varphi = \arccos(m) \tag{71}$$

(see Fig. 5). The probability that an input pattern $\tilde{\xi}$ yields the right response is the probability $\rho(\tilde{\xi})$ that :

$$(\tilde{J}^T.\tilde{\xi})(\tilde{J}^P.\tilde{\xi}) > 0$$

This probability is given by the relative volume which is determined by the two planes $\tilde{J}^T.\tilde{\xi} = 0$ and $\tilde{J}^P.\tilde{\xi} = 0$ in the N-dimensional hypersphere. Therefore :

$$\rho(\tilde{\xi}) = 1 - \frac{\varphi}{\pi}$$

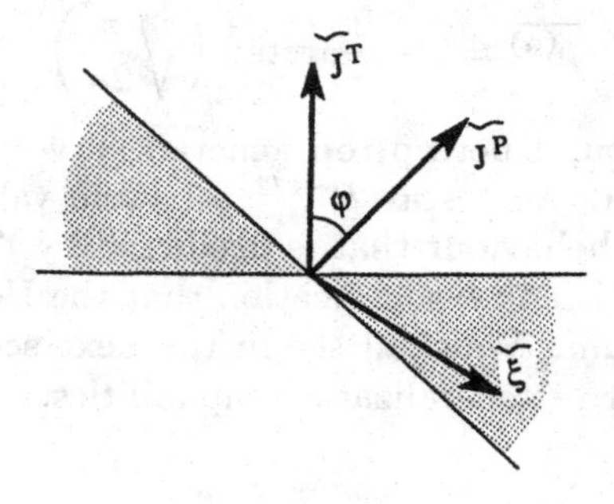

Figure 5

The probability that a particular pupil perceptron with overlap m with the teacher perceptron produces convenient responses for $P = \alpha N$ patterns is :

$$\rho^{(P)} = \left(1 - \frac{\varphi}{\pi}\right)^P$$

The number of perceptrons with overlap m is :

$$\mathcal{N}(m) = \binom{N}{mN} = \frac{N!}{\left(N\frac{1+m}{2}\right)!\left(N\frac{1-m}{2}\right)!} \tag{72}$$

and therefore the number of convenient pupil machines whose overlap with the teacher machine is m is :

$$\Gamma(m) = \left(1 - \frac{\varphi}{\pi}\right)^P \frac{N!}{\left(N\frac{1+m}{2}\right)!\left(N\frac{1-m}{2}\right)!}$$

With the Stirling formula :

$$\mathrm{Log}(n!) \simeq n(\mathrm{Log}(n) - 1)$$

this number is written as :

$$\Gamma(m) \simeq \exp\left[N\left\{\alpha\mathrm{Log}\left(1 - \frac{1}{\pi}\arccos(m)\right) - \cdots \right.\right.$$
$$\left.\left. -\frac{1}{2}\left((1+m)\mathrm{Log}(1+m) + (1-m)\mathrm{Log}(1-m) - 2\mathrm{Log}2\right)\right\}\right]$$

For small enough α, the exponent

$$F(\alpha, m) = \alpha\mathrm{Log}\left(1 - \frac{1}{\pi}\arccos(m)\right) - \cdots$$
$$-\frac{1}{2}\left((1+m)\mathrm{Log}(1+m) + (1-m)\mathrm{Log}(1-m) - 2\mathrm{Log}2\right) \tag{73}$$

vanishes meaning that no more than one machine with overlap m is able at responding correctly to the set of P patterns. The critical value is therefore given by :

$$\alpha = \frac{\left((1+m)\mathrm{Log}(1+m) + (1-m)\mathrm{Log}(1-m) - 2\mathrm{Log}2\right)}{2\mathrm{Log}\left(1 - \frac{1}{\pi}\arccos(m)\right)}$$

The maximum value α_c over m is :

$$\alpha_c = 1.448\cdots$$

Derrida and Gardner argue that below α_c there may exist different pupil perceptrons which give the correct answers for the training set. Above this value one is certain that the only possible machine is a replica of the teacher perceptron $\tilde{J}^P \equiv \tilde{J}^T$. α_c is therefore an upper limit for this identification to occur.

4.5. A STATISTICAL MECHANICS APPROACH

The theories of Gardner and Derrida and that of Vallet rest upon combinatorial arguments. The problem of binary perceptrons has been reconsidered by Sompolinsky, Tishby and Seung [19] in a spirit which is closer to that of statistical mechanics.

These authors use error based cost functions and the limit $T = 0$. Therefore the cost $H(\tilde{J}^P)$ that is attached to every point $\tilde{J}^P$ of the phase space of the system is given by :

$$H(\tilde{J}^P) = \sum_{\mu=1}^{\alpha N} \theta\left[-(\tilde{J}^T.\tilde{\xi}^\mu)(\tilde{J}^P.\tilde{\xi}^\mu)\right] \tag{74}$$

The Metropolis algorithm with eq. (74) as its cost function creates an ensemble of pupil perceptrons in thermal equilibrium :

$$\rho(\tilde{J}^P) = \frac{1}{Z}\exp[-\frac{H(\tilde{J}^P)}{T'}] \tag{75}$$

with

$$Z = \sum_{\tilde{J}^P} \exp[-\frac{H(\tilde{J}^P)}{T'}]$$

The generalization error is defined by :

$$H^G(\tilde{J}^P) = \frac{1}{2^N} \sum_{\text{All } \{\tilde{\xi}\}} \theta\left[-(\tilde{J}^T.\tilde{\xi})(\tilde{J}^P.\tilde{\xi})\right]$$

where $\tilde{\xi}$ spans the 2^N possible input states.

We are interested in average quantities, namely quantities not depending on the particular training set $\mathcal{E}$ we have chosen. Therefore a sample average over training sets has to be carried out. This involves appealing to the replica technique that we have already mentioned. Calculations will not be detailed here. Sompolinsky et al. [19] show that, in the limit of large T's, the distribution (75) may be reduced to :

$$\rho(\tilde{J}^P) \simeq \frac{1}{Z^G}\exp[-\frac{P}{N}\frac{H^G(\tilde{J}^P)}{T'}] \; ; \; Z^G = \sum_{\tilde{J}^P} \exp[-\frac{P}{N}\frac{H^G(\tilde{J}^P)}{T'}]$$

and the sample averaged generalization error to :

$$F^G = -T'\text{Log}(Z^G)$$

The computation of Z^G is achieved in two steps as it is customary in statistical physics. A first sum is carried out over all $\tilde{J}^P$ such as $\tilde{J}^P.\tilde{J}^T = Nm$. This provides the average generalization error $f^G(m)$ per link :

$$Z^G = \sum_m \sum_{\tilde{J}^P(m \text{ given})} \exp[-\frac{P}{N}\frac{H^G(\tilde{J}^P)}{T'}]$$
$$= \sum_m \exp[-\frac{N f^G(m)}{T'}]$$

One has :

$$f^G(m) = \alpha u(m) - Ts(m)$$

where $u(m)$ is the average error for an input pattern $\tilde{\xi}$ feeding a pupil perceptron with overlap m and $s(m)$ is the entropy, the number of pupil perceptrons with overlap m. Using the argument leading to eq. (71) one finds :

$$u(m) = \frac{\varphi}{\pi} = \frac{1}{\pi}\arccos(m)$$

and $s(m)$ is given by eq. (71). Finally :

$$\frac{1}{T'} f^G(m) = \frac{\alpha}{T'\pi}\arccos(m) + \frac{1}{2}\left[(1-m)\text{Log}(1-m) + (1+m)\text{Log}(1+m)\right] + \text{Log}2$$

The second step consists in a steepest descent in order to finding out the overlap which minimizes the generalization error. m_0 is the solution of :

$$\frac{df^G(m)}{dm} = 0$$

leading to :

$$m_0 = \tanh\left[\frac{\alpha}{T'\pi}(1-m_0^2)^{1/2}\right]$$

$m = 1$ is always a minimum of $f^G(m)$ (Fig. 6) : There always exists a pupil machine which reproduces the behaviour of the teacher which is simply a copy of the teacher.

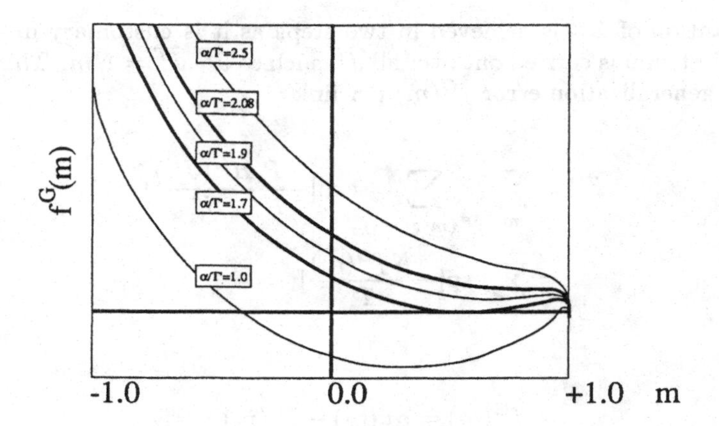

Figure 6. Average error for various values of parameter α/T'.

For $\alpha/T' < 2.08$ there is another minimum m_0 with $m_0 < 1$, which means that it is less stable than the other minimum $m = 1$, for $1.70 < \alpha/T' < 2.08$. For $\alpha/T' < 1.70$ the minimum at m_0 is stable. In terms of machines for $\alpha/T' < 1.70$ the Metropolis dynamics converges always towards machines which generalize poorly (since $m_0 < 1$). For $1.70 < \alpha/T' < 2.08$ the target of the dynamics is the teacher perceptron which is the most stable solution, but as the metastable states are so numerous the dynamics first selects poorly generalizing machines. One has to wait an exponentially long time before the dynamics finds the correct machine. For $\alpha/T' > 2.08$ the dynamics converges fast to the teacher perceptron.

Many more details can be found in the quoted article. For example it is shown that the property that perfect generalization is achieved for a finite number of training patterns is conserved at all training temperatures. This property is also observed for perceptrons that are not rigorously binary, that is to say when the weights are no more strictly constrained to $J_i \in \{+1, -1\}$ but are permitted to take values in the vicinity of $+1$ or -1.

5. Networks with Hidden Units

5.1. LEARNING ALGORITHMS AND GENERALIZATION IN FEED-FORWARD NETWORKS

Braitenberg says that the cortex is an organ which mainly speaks to itself. The optical nerve of man, for example is made of about one million fibers whereas the visual cortex comprises about one billion cells. Therefore the problem of learning in system with hidden units cannot be ignored. This problem is that of internal representations, that is to say that of determining the set of states $I^{h,\mu}$ of hidden units which are to be associated to states $I^{v,\mu}$ of visible units. If solved, learning reduces to learning in visible networks. To date the problem of internal representations has only been considered for feed-forward networks. Let us summarize the various ways the problem has been tackled.

a) In the Neocognitron of Fukushima, the internal representations are directly given (by hand so to speak) to the various layers of the network. Learning reduces to perceptron learning.

b) In the Back propagation algorithm of Rumelhart, Le Cun, Parker, [12,18] the internal representations are determined by the neuronal dynamics which may seem quite natural. Since there is no cost function which reproduces the neuronal dynamics eq. (22) in feed-forward network, it is necessary to introduce this dynamics as a *constraint* that the system has to satisfy. The patterns have then to be learned one by one and, for each pattern $I^{v,\mu}$ the cost that the system has to minimize is :

$$H(I^{\mu},\{J\}) = \sum_{i=0}^{\mathcal{N}_O} \theta\left[-\gamma_i^{\mu}\right] + \sum_{i=0}^{\mathcal{N}_h} \lambda_i \left[\sigma_i - \mathrm{Sgn}(h_i^{\mu})\right]$$

where

$$\gamma_{i\in O}^{\mu} = \xi_i^{\mu} \sum_j J_{ij}\sigma_j$$

j belonging to the next to last layer, h_i^{μ} is the field observed on hidden site i when the input state is $I^{v,\mu}$ and λ_i is a Lagrange parameter. The parameters $\{J\}$ are determined by a gradient algorithm with :

$$\frac{\partial H}{\partial \lambda_i} = 0 \ ; \quad \frac{\partial H}{\partial \sigma_i} = 0 \ \text{ and } \ \Delta\tilde{J} = -\varepsilon \frac{\partial H}{\partial \tilde{J}}$$

Since the function $\mathrm{Sgn}(x)$ is non-analytical which hinders the derivatives to be computed, it is replaced by a smoother sigmoidal function $\mathcal{S}(\beta x)$. $\mathcal{S}(\beta x)$ is called the response neuronal function. One observes that this algorithm is not a minimax procedure and that learning is serial.

c) The algorithm of learning by Choice of internal representations (Chir) of Domany and Meir [7] consists in handling the internal states at will so as to make them apt at being learned when using perceptron algorithms regardless of the neuronal dynamics. The idea is that the neuronal dynamics aims at lowering a "local cost" as does the perceptron algorithms and therefore that optimizing the perceptron costs results in a good performing network when using the neuronal dynamics. Here learning can be made parallel.

d) Mezard and Nadal [14] have put forward a learning algorithm, called the Tiling algorithm, which is constructive in nature. Learning builds the network layer after layer until the system associates the right answers with all inputs of the training set $\mathcal{E}$. Each step consists in training the units of a layer by using perceptron learning algorithms and adding units to the layer until a faithful internal representation of the training set is obtained for this layer. An internal representation is faithful if it does not contain identical states associated with different outputs. A theorem, which states that the number of errors is a decreasing function of the number of layers, makes sure that the algorithm converges towards a convenient solution, a machine which produces all expected responses of the training set.

e) Finally, pruning algorithms inspired by the general idea of Darwinian learning (Changeux [3]), have also been imagined (Dehaene, Peretto). Learning proceeds in two step : A first phase consists in a random sprouting of synaptic contacts giving rise to many sorts of circuits. The second phase is that of pruning where experience (the training set) selects among the avalaible circuits the one which fits the desired response.

We now give a few comments as regards the way these algorithms behave with respect to generalization. It is reminded that generalization demands two sorts of restrictions to be effective. On the one hand it is necessary that the training set of patterns may be generated by using simple networks. On the other hand the choice of possible architectures must be somehow limited to a restricted set of architectures.

Simulations carried out by using of the Back propagation algorithm have shown that too small or too large networks are not apt at generalizing, the former sort of network because it is not even able to learn the training set correctly, the latter because it learns the training set too easily. That is to say it learns "by heart" all examples of the set without trying to extract some generating rule from the set. Therefore there must exist some optimal size that allows the network to generalize at best. However this size is problem dependent and it is hard to determine *a priori* what best size the network should take. A strategy advocated by Le Cun and Denker is starting from large enough networks and introducing more and more severe constraints afterwards. An example of constraint is forcing some weights to be one another identical for example which amounts to restricting the number of possible architectures. We have the same type of problem with the Chir algorithm

The effect of constraints is more transparent in Darwinian learning. During the sprouting phase there is a certain probability for a given function to be realized. Generalization means selecting a given function among the available functions. Either the system easily generates the desired function and a few examples will be enough during the pruning phase to extract the function : Generalization is achieved. If the case does not arise the system is unable of generalizing the particular task that is embedded in the training set.

Let us assume that the training set is generated by some simple rule, that is to say by a rule which may be materialized by a teacher network comprising a number of hidden units which is not too large as compared with the number of visible units. On the other hand let an architecture be given such as the volume Γ^R of machines which reproduce all possible examples generated by the rule, does not vanish. The volume Γ^P of solutions for pupil machines which reproduce the examples of a training set $\mathcal{E}$ necessarily includes Γ^R, $\Gamma^P \supseteq \Gamma^R$. When the number P of examples increases the volume Γ^P shrinks and generalization is achieved if $\Gamma^P = \Gamma^R$. However if the phase space of the network parameters is very large the decrease may be so slow that all possible patterns have to be learned before the merging of the two volumes Γ^P and Γ^R occurs. These considerations suggest a general strategy : Let us assume that one is able to find the minimal network, that is to say the network involving the smallest number of parameters such as $\Gamma^P \neq 0$. This volume may be tiny. Eventually it is reduced to a single point. Probably for this number P of training examples there does not exist any solution that satisfies all examples of the rule and $\Gamma^R = 0$. However let P grow. For every increasing P one is compelled to introduce larger and larger minimal architectures but the volume of solutions which satisfies the examples of the set still remains reduced to, say, one point. If the underlying rule is simple, there must exist some low enough value of P such as $\Gamma^R \neq 0$. Since Γ^P is small and $\Gamma^P \supseteq \Gamma^R$ one must have $\Gamma^P = \Gamma^R$ and generalization is achieved.

This is the reason why constructive learning algorithms sush as the one that has been put forward by Mezard and Nadal are so interesting : If minimal networks are found at every step of the learning stage one is sure to find a network which generalizes at best. This point has been emphasized by Anshelevich et al. [1] in particular.

5.2. SOME PRELIMINARY RESULTS ON GENERALIZATION CAPABILITIES OF FEED-FORWARD NEURAL NETWORKS

The problem of generalization in layered and in more general networks is under current scrutiny. A few results have been obtained so far.

a) The first statement is that it is always possible to realize any Boolean function which maps N binary input units onto a unique output unit by using a three layer feed-forward network defined by :

$$\mathcal{N}_I = N \quad ; \quad \mathcal{N}_H \quad ; \quad \mathcal{N}_O = 1$$

and

$$\mathcal{N}_H \leq 2^{N-1}$$

The solution is a *grand mother solution* where one manages to have only one hidden unit $h(\mu)$ in state $\sigma_{h(\mu)} = +1$ and all other units in state $\sigma_h = -1,\ h \neq h(\mu)$ for all P patterns which must produce a $\xi^\mu = +1$ ($P \leq 2^{N-1}$). This may be realized by choosing the weights impinging on the hidden units as :

$$J_{h \in H, i \in I} = \xi_i^\mu \text{ for } h = h(\mu)$$

and the threshold as :

$$J_{h,0} = -N + 1$$

To obtain the desired output it is enough to make :

$$J_{o \in O} = +1$$

with a threshold :

$$J_0 = P - 1$$

Therefore it is always possible to solve the problem of implementing a Boolean function of N entries by using an exponential number of neurons. One knows that the problem of finding the combination of basic logical gates that produces a given Boolean mapping is a hard problem. It is even the archetypical NP-complete problem and, as far as one knows for the moment, an NP-complete problem demands an exponential number of steps to be solved. One sees that the problem of generalization in neural networks is very close to the problem of algorithmic complexity with the resolution time replaced by the amount of units that is needed to solve the problem.

b) It has been shown, by Judd [9] in particular, that
- finding the set of functions which a particular architecture is able to produce
- finding the minimal network which implements a given function

are hard (NP-complete) problems.

To escape this pitfall it is necessary to have some information regarding the type of architecture which produces the examples of the training set. For example one may assume that the function which is to be learned has been produced by a perceptron. The family of functions which a perceptron can produce is called the family of linearly separable functions. The total number of Boolean functions of N argumens is :

$$\mathcal{N}_B = 2^{2^N}$$

a huge number indeed, whereas the number of linearly separable functions is :

$$\mathcal{N}_{\mathcal{L}} \simeq \frac{2^{N^2}}{N!}$$

a still very large number. The number $\mathcal{N}_{\mathcal{L},B}$ of functions that are linearly separated by binary perceptrons is lower. It is given by the number of possible machines :

$$\mathcal{N}_{\mathcal{L},B} = 2^N$$

However a perceptron is not able to learn more than $2N$ random patterns (0.83N for binary perceptrons). Therefore if one strives to teach a pupil perceptron a number P of patterns produced by a teacher perceptron which is larger than this value the constraint on the weights of the pupil perceptron will be so strong that there will be only one solution left : The pupil produces a correct answer for all possible inputs, that is to say the pupil machine generalizes for a linear number $P \simeq N$of examples. In other words learning with $P \simeq N$ will be enough to determine one Boolean function among the $\mathcal{N}_{\mathcal{L}}$ possible linearly separable Boolean functions. One sees how important is the fact that the architecture of the teacher machine is known.

It is then natural to investigate how the storage capacity of random patterns changes when the architecture changes. Partial results are available.

c) A k-perceptron problem has been studied by Mezard and Patarnello [15]. In this problem one has a number k of perceptrons all sharing the same N input units. The k output units of the perceptrons are inputs of some *well defined Boolean function* which maps the k states onto a unique output. An example of a 2-perceptron is a network whose N inputs feed two perceptrons a and b. The two output units of the perceptrons makes a hidden layer of two elements which are inputs of a XOR Boolean function (for example). The problem is finding the maximum number P of random patterns which can be learned by the system. A simple reasoning would be the following : A given output ξ^μ may be produced by two states of the hidden layer. For example $\xi^\mu = +1$ is the result of either $\xi^\mu_a = -\xi^\mu_b = +1$ or $\xi^\mu_a = -\xi^\mu_b = -1$. Therefore the number of storable patterns is twice the number of patterns that can be memorized in a single perceptron, namely $P_c = 4N$. This is not correct.

Let us call $I^{\mu+}$ the patterns of the training set $\mathcal{E}$ which produce an output $\xi^\mu = +1$ and $I^{\mu-}$ the patterns which produce an output $\xi^\mu = -1$. Given a set of connections $\{J\}$, a pattern of the first category yields a convenient response if the fields h^μ_a and h^μ_b it produces on the two perceptrons have opposite signs. Therefore as far as the first family of patterns is concerned the point $\{J\}$ of the phase space of interactions belongs to the volume of solutions if :

$$\prod_{\mu+} [\theta(h_a^\mu)\theta(-h_b^\mu) + \theta(-h_a^\mu)\theta(h_b^\mu)] = 1 \tag{76}$$

Otherwise this quantity vanishes. In actual fact for $\{J\}$ to be a solution a condition similar to eq. (76) must also be obeyed for the family $I^{\mu-}$. As a whole the number of solutions is given by :

$$\Gamma = \sum_{\{J\}} \prod_{\mu+} [\theta(h_a^\mu)\theta(-h_b^\mu) + \theta(-h_a^\mu)\theta(h_b^\mu)] \prod_{\mu-} [\theta(h_a^\mu)\theta(h_b^\mu) + \theta(-h_a^\mu)\theta(-h_b^\mu)] \tag{77}$$

From a technical point of view it is necessary to introduce a constraint on the space that is avalaible to $\{J\}$ as was done in eq. (60) and finally the quantity of interest is :

$$\begin{gathered} \Gamma = \int \prod_{a,j} dJ_{aj} \prod_{a} \delta(\sum_j J_{aj}^2 - N) \prod_{a,\mu} \delta(h_a^\mu - \sum_j J_{aj}\xi_j^\mu) \cdots \\ \prod_{\mu+} [\theta(h_a^\mu)\theta(-h_b^\mu) + \theta(-h_a^\mu)\theta(h_b^\mu)] \prod_{\mu-} [\theta(h_a^\mu)\theta(h_b^\mu) + \theta(-h_a^\mu)\theta(-h_b^\mu)] \end{gathered} \tag{78}$$

The calculation is carried out by using the replica technique along the lines pioneered by E. GARDNER. Assuming that the number of $I^{\mu+}$ patterns is the same as that of $I^{\mu-}$ patterns, the critical value $\alpha_c = P_c/N$ is now given by :

$$\alpha_c = \frac{1}{\int_0^{+\infty} \frac{dy}{\sqrt{2\pi}} y^2 \exp\left(-\frac{y^2}{2}\right) \left[2 \int_y^{+\infty} \frac{dx}{\sqrt{2\pi}} \exp\left(-\frac{x^2}{2}\right)\right]} = \frac{4}{1 - 2/\pi} \simeq 11 \tag{79}$$

which is sensitively larger than the storage capacity $\alpha_c = 4$ which could have been guessed. MEZARD and PATARNELLO have also carried out the computation of the storage capacity of general k-perceptrons. The formula (78) becomes :

$$\begin{gathered} \Gamma = \int \prod_{a,j} dJ_{aj} \prod_{a} \delta(\sum_j J_{aj}^2 - N) \prod_{a,\mu} \delta(h_a^\mu - \sum_j J_{aj}\xi_j^\mu) \cdots \\ \prod_{\mu+} [\sum_{I^{h+}} \prod \theta(\xi_a^{h+} h_a^\mu] \prod_{\mu-} [\sum_{I^{h-}} \prod \theta(\xi_a^{h-} h_a^\mu] \end{gathered} \tag{80}$$

where I^{h+}, (resp. I^{h-}), is a state of the phase space of hidden units which yields a positive (resp. negative) response on the output unit and ξ_a^{h+}, (resp. ξ_a^{h-}), the state of hidden unit a for hidden state I^{h+}, (resp. I^{h-}). The storage capacity of the k-perceptron is given by :

$$\alpha_c = \frac{1}{\int_0^{+\infty} \frac{dy}{\sqrt{2\pi}} y^2 \exp\left(-\frac{y^2}{2}\right) \left[2 \int_y^{+\infty} \frac{dx}{\sqrt{2\pi}} \exp\left(-\frac{x^2}{2}\right)\right]^{k-1}} \tag{81}$$

α_c diverges as $k \to \infty$ like $\frac{2}{\pi}k^3$. This is a very surprising result since the number of connections is kN. The only way out is that a connection carries an average of $\simeq \frac{2}{\pi}k^2$ bits of information. If this statement is right the capacity must be sensitive to slight perturbations of synaptic efficacies which is at odds with the usually accepted robustness of neural networks. In actual fact doubts have arosen concerning the validity of these results. A lower bound of the storage capacity per link would scale as Log(k) instead of k^2 (Mitchinson et al.). Simulations seems to confirm this prediction (Parga). What is probably uncorrect in the Mezard derivation is the hypothesis of symmetrical replica.

d) The calculation of Mezard and Patarnello is, so to speak, an in breadth exploration of layered networks. We have carried out a complementary study, an exploration in depth of a network that implements the classical Boolean gates, that is to say a network with two inputs and one output. There are sixteen gates and the problem is that of studying the probability for a particular gate to be realized when the set $\{J\}$ of parameters is chosen at random.

The phase space of input units is made up of four states I^μ, $\mu = 1, \cdots 4$. We consider the set of states $\sigma_j^h(I^\mu)$ of unit j of hidden layer h when the input state is I^μ. The four states of σ_j^h corresponding to the four inputs determines one of the 16 possible Boolean functions. Therefore the units of layer h may be categorized according to the function they display. Let us then define $\tau_n^\mu \in \{+1, -1\}$, with $\mu = 1, \cdots, 4$, $n = 1, \cdots, 16$ as the ouput of the Boolean gate n for input μ. The equation :

$$\sigma_i^{h+1}(I^\mu) = \text{Sgn}\left[\frac{1}{\sqrt{\mathcal{N}_H}} \sum_{j=0}^{\mathcal{N}_H} J_{ij}^h \sigma_j^h(I^\mu)\right] \tag{82}$$

may be replaced by :

$$\sigma_i^{h+1}(I^\mu) = \text{Sgn}\left[\sum_{n=1}^{16} K_n^h \tau_n^\mu\right] \tag{83}$$

where

$$K_n^h = \frac{1}{\sqrt{\mathcal{N}_H}} \sum_{j/\sigma_j^h(I^\mu) \equiv \tau_n^\mu,\ \forall\mu} J_{ij}^h$$

is a sum over all neurons of layer h which implement a given Boolean gate n. Now the unit i of layer $h+1$ implements a Boolean gate m if :

$$\prod_{\mu=1}^{4} \theta[\tau_m^\mu \sum_{n=1}^{16} K_n^h \tau_n^\mu] = 1 \tag{84}$$

Otherwise this quantity vanishes. The effective interactions K_h^n are random variables whose distributions are Gaussian and given by :

$$\rho(K_n^h) = \frac{1}{\sqrt{2\pi\Delta^2\rho_n^h}} \exp[-\frac{(K_n^h)^2}{2\Delta^2\rho_n^h}]$$

if the interactions J_{ij} are symmetrically distributed with variance Δ. In this formula we have used the fact that, on average, the number of neurons which implement a Boolean gate n is proportional to the probability ρ_n^h for the function n to be implemented in that layer h. Then :

$$\begin{aligned} \rho_m^{h+1} &= \sum_{\{K^h\}} \prod_{n=1}^{16} \rho(K_n^h) \prod_{\mu=1}^{4} \theta[\tau_m^\mu \sum_{n=1}^{16} K_n^h \tau_n^\mu] \\ &= \int \prod_{n=1}^{16} dK_n^h \frac{1}{\sqrt{2\pi\Delta^2\rho_n^h}} \exp[-\frac{(K_n^h)^2}{2\Delta^2\rho_n^h}] \prod_{\mu=1}^{4} \theta[\tau_m^\mu \sum_{n=1}^{16} K_n^h \tau_n^\mu] \end{aligned} \tag{85}$$

This equation is solved by exponentiating the step function $\theta(x)$. One finds recursion relations which couple the probabilities of finding a Boolean gate m on one site of layer $h+1$ given the probabilities of finding the gates on a site of the preceding layer. In principle we are faced with a set of sixteen coupled equations. Fortunately symmetry considerations show that some gates have the same probabilities which allow to make them categorized along 4 families of Boolean gates. The families are called V (2 members), D (4 members), A (8 members) and X (2 members). If the ordering of the 4 inputs is $(+,+)$, $(+,-)$, $(-,-)$, $(-,+)$, the corresponding outputs, $(+,+,+,+)$ is an example of τ_V, $(+,+,-,-)$ an example of τ_D, $AND \equiv (+,-,-,-)$ an example of τ_A and $XOR \equiv (+,-,+,-)$ an example of τ_X. Therefore it is enough to know the four probabilities ρ_V^0, ρ_D^0, ρ_A^0 and ρ_X^0 to determine all other probabilities. The probabilities ρ^0 are derived from calculations of volumes of the 3-dimensional cube spanned by J_{00} (the threshold), J_{10} and J_{20} which correspond to each family. One finds :

$$\rho_V^0 = \rho_D^0 = \frac{1}{12} = 0.0833$$

$$\rho_A^0 = \frac{1}{16} = 0.0625$$

$$\rho_X^0 = 0$$

which means, in particular, that there is no perceptron apt at implementing the XOR function. Introducing one hidden layer modifies theses probabilities. The probability of finding a XOR gate does not vanish any more (Fig. 7). However it remains significantly lower than the other probabilities (even in the limit of large $\mathcal{N}_{H=1}$). The analytical derivation yields :

$\rho_V^0 = 0.0783$, $\rho_D^0 = 0.0786$, $\rho_A^0 = 0.0597$, $\rho_X^0 = 0.0253$

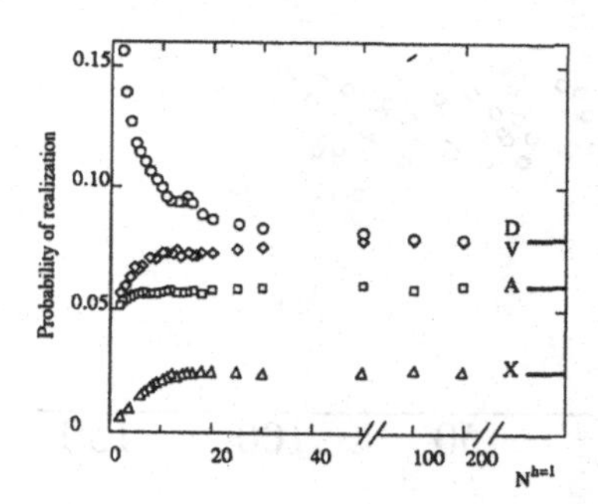

Figure 7. Boolean gates distribution for a $2-\mathcal{N}_{\rm H}-1$ layered network as a function of $\mathcal{N}_{\rm H}$ the number of hidden units.

The distribution of Boolean gates becomes uniform when the number of hidden layers increases. Convergences to the common value $\rho^\infty = 1/16 = 0.0625$ is achieved within 10% for $H = 5$.

e) Finally one would like to obtain some information regarding the generalization properties of more general feed-forward nets.

Implementing successfully any, unrestricted Boolean function in a network means that one is unable to foresee what the output of the system should be by changing one single state of the input units. This is not so in perceptrons. Let us consider a binary perceptron whose connections $J_i \in \{-1, +1\}$, $i = 1, \cdots, N$ are chosen at random and the output ξ^μ it produces for an input $I^{i,\mu}$. The output may change only in case that the field :

$$h(I^\mu) = \sum_{i=1}^{N} J_i \xi_i^\mu$$

on the output unit is (or close to) zero. The probability for this to happen is $\rho^0 = 1/\sqrt{N}$. One could say that linearly separable functions, the functions that can be realized by perceptrons, show a sort of continuity (rigidity) property and that this property, in turn, makes perceptrons unable to implement any kind of Boolean function. It is therefore interesting to look at the probability

$\rho^h(\{\mathcal{N}_H\})$ that the output of a layered network with h hidden layers, each comprised of $\mathcal{N}_H$ units, changes after the reversal of one single unit in the input layer.

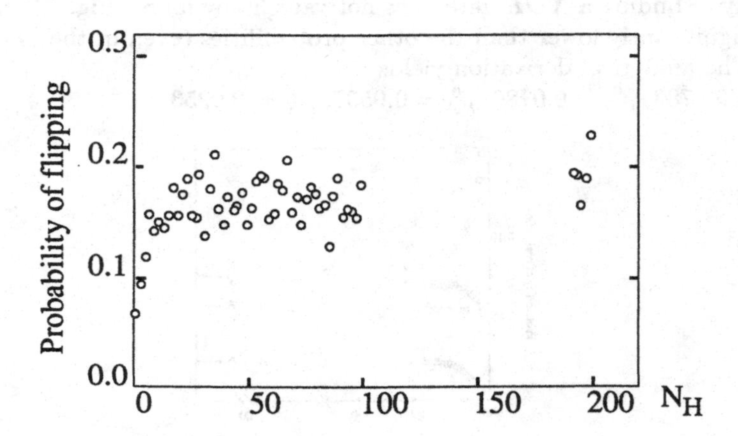

Figure 8. Probability of modifying the state of the ouput unit of a $\mathcal{N}_\mathrm{I}$–$\mathcal{N}_\mathrm{H}$ – 1 layered network by flipping one randomly chosen input unit as a function of $\mathcal{N}_\mathrm{H}$. $\mathcal{N}_\mathrm{I}$= 99.

For systems with one hidden layer simulations show that this probability increases as the number of neurons of the hidden layer increases but it saturates to a value well below 1/2 indicating that enlarging the number $\mathcal{N}_1$ of hidden units is not enough to make networks apt at realizing any Boolean function at least as far as $\mathcal{N}_I$ is not exponentially large. One finds $\rho^{h=1}(\infty) = 0.16 \mp 0.02$ (for $\mathcal{N}_I = 99$) (Fig. 8).

The same result is observed when layers are added, with progressively increasing saturation values. As in the case of Boolean gates, saturation to the limit $\rho^h = 1/2$ occurs for $h \simeq 5$. The conclusion is that networks comprised of less than 5 layers show significant rigidity. If rigidity is a necessary condition for networks to generalize these simulations suggest that the number of layers of generalizing feed-forward networks must be less than 5. It must be realized that the condition is sufficient to be sure that the network is unable to implement any Boolean function. By no means has it been shown that the condition is necessary. It is all the more likely that larger networks still remain unable to implement general Boolean functions. If the case arises it simply means that rigidity is not enough to characterize families of Boolean functions.

5.3. RECURSIVE NETWORKS WITH HIDDEN UNITS

This section will be short since practically no analytical result is known concerning the generalization capabilities of most general networks in spite of the fact that this problem is central in the understanding of the cortical properties for example.

Let $\mathcal{N}_H$ be the number of hidden units and $\mathcal{N}_V$ that of visible units. The network made up of the sole visible units may store a maximum of $2 \times \mathcal{N}_V$ random patterns. The idea is that if the number of hidden units is large enough, such as $\mathcal{N}_H \simeq 2^{\mathcal{N}_V}$ for example, it may be possible to associate one input pattern to one internal state so driving apart even close visible patterns. Making $\mathcal{N}_H$ decrease forces certain internal states to merge so giving rise to generalization properties. However, this nice program comes up against a major problem, that of ***obsession***. To see where obsession comes from one considers the field created by the visible units on hidden unit i for input pattern I^μ :

$$h_{i\in H}(I^\mu) = \sum_{j\in V} J_{ij}\xi_j^\mu$$

We assume that the weights J_{ij} are weak and random. The state of unit i is then given by :

$$\sigma_i^\mu = \text{Sgn}(h_i(I^\mu))$$

The weights are modified and if one assumes that the learning dynamics is Hebbian, then :

$$J_{ij} \simeq \sigma_i^\mu \xi_j^\mu$$

The system now experiences a second pattern. The field becomes :

$$h_i(I^{\mu'}) = \sum_j J_{ij}\xi_j^{\mu'} \simeq \sigma_i^\mu \sum_j \xi_j^\mu \xi_j^{\mu'} = \sigma_i^\mu \left(I^\mu . I^{\mu'}\right)$$

Therefore the fields $h_i(I^{\mu'})$ are parallel to the fields $h_i(I^\mu)$. The hidden state $\{\sigma_{i\in H}\}$ for $I^{\mu'}$ is identical (or opposite) to I^μ. If Hebbian learning continues the internal representation of the network is made up of a unique state. This situation appears for feed-forward networks as it appears for recursive networks. For the latter, however, the situation is even worse since contributions of other hidden units are to be added to the field. Due to the field of visible units the state σ_i of hidden units tends to be identical to σ_i^μ and therefore :

$$h_{i\in H}(I^{\mu'}) = \sum_{j\in V} J_{ij}\xi_j^{\mu'} + \sum_{j\in H} J_{ij}\sigma_j^\mu = \sigma_i^\mu \left(I^\mu . I^{\mu'} + \mathcal{N}_H\right)$$

As far as feed-forward nets are concerned, avoiding the collapse of internal states is precisely what the tiling algorithm strives to achieve thanks to the notion of faithfulness. The same constraint is also present in the Chir algorithm although less clearly stated. Finally back propagating errors is the way the back-propagation algorithm uses to avoid the merging of internal states.

How then efficient learning (and generalization) may be realized in recursive networks?

A general theory of learning using the statistical methods we have introduced in section 2 has not been worked out for these systems yet. In principle the Boltzmann machine should be the answer. However this algorithm involves relaxation stages and the problem of obsession remains. In actual fact for the Boltzmann machine to be an efficient algorithm it is necessary that the number of hidden units remains low as compared with that of visible units.

A way out is to put constraints on connections so as to reduce a general recursive net to one which is more open to efficient learnings. Two possibilities can be put forward :

- Either one freezes the internal hidden-hidden (H-H) connections and learning is achieved by modifying the visible-hidden (V-H) connections.
- Or one freezes the (V-H) connections and learning is achieved by modifying the (H-H) connections.

a) The first possibility is called ***self-organization.*** The internal (H-H) connexions do exist but they cannot be modified. These connections must provide a convenient, large enough, set of stable internal states. A way of achieving large internal degeneracy is to use a frustrated set of connections. For example lateral inhibition provides a mean of creating such frustrated networks. The other (H-V) connections are plastic and, in principle, any learning algorithms that have been imagined for feed-forward nets could be used. The resulting learning algorithms form the essence of self-organizing dynamics of Kohonen [10]. They display in many instances good generalization capabilities by grouping together input that share some similarities. For example the Kohonen algorithm is able to find out the families of phonemes of the Finish language that are commonly distinguished by classical phonetical analysis.

b) The other possibility has not been exploited yet. It simply consists in creating random (V-H) connections. These connections are non plastic. One assumes, moreover, that the overall activity of input cells is large during learning $\sigma^{\mu}_{i\in V} = \lambda\xi^{\mu}_{i}$ (large λ), and weak during the relaxation of the system (small λ). Then a hidden state $\{\sigma^{\mu}_{i\in H}\}$ is associated with each input states I^{μ} and one may consider the network of hidden units as a visible network at least during the learning stage. The (H-H) connections are modidied according to some perceptron algorithm accordingly.

6. Conclusion

This paper is manifestly open ended. It stresses the fact that the problem of learning has only been partially solved. Convenient training dynamics have been found for visible networks. The situation has favourably evolved for feed forward networks but one still ignores how to efficiently train a general recursive network. The related problem of generalization is even less understood although generalization is the property that makes neural networks so appealing. Some results have been found for visible networks. However, what is known on feed-forward networks still remains very sparse. In many instances only empirical data, obtained by using simulations, are avalaible. Eventhough one does not really understand the reasons of the successes the results have been often encouraging. They prompt to carry out further investigations in that domain.

We end this survey with a small digression. Generalization poses a series of deep problems which concern the relationship between a machine and its environment. We have seen that generalization is a meaningful notion if, on the one hand, the data that are provided by the environment are not quite impredictible and, on the other hand, if there exist constraints that the architecture of the machine has to obey. Both conditions are necessary.

In general, the training set is a random subset of all possible patterns that environment may create. When the system does not control the set of patterns it is trained to one says that learning is passive. One can imagine a more efficient way of learning, namely active learning. Active learning is suggested by the way experimental psychology defines generalization. For a psychologist an animal generalizes if it is trained to respond to a given stimulus (learning stage) and if it yields the same response for stimuli it has not been trained to but that are close to the training stimulus. For example a dog is trained to respond to a tune with given frequency. After training the dog responds to sounds with close frequencies. This is the sort of generalization we had in mind when discussing the "rigidities" of Boolean functions. However the dog can also be trained to discriminate between sound with close frequencies. For example it can be trained to respond positively for a certain frequency and negatively to another frequency which is close to the former. Animals discriminates very efficiently which means that there exists a certain mechanism which enhances the learning performance as soon as the experienced stimulus is not that which is expected, when there is a surprise. The idea is then that rather trying to learn random patterns it could be more efficient to learn a selected set of patterns, those that are surprising. As the environment does not automatically select the interesting patterns, the system has to choose a strategy, to make experiments, to ask questions to the environment, so as to select the most interesting patterns and to learn them afterwards. This is active learning and certainly a good way for generalizing. Let us give an example :

One considers the problem of a pupil binary perceptron (the machine) which strives to learn the function that is generated by a teacher binary perceptron (the environment). This problem contains the two ingredients that are necessary for the system to generalize. The architecture of the machine is defined and the environmental function is simple. We have seen how this system generalizes passively. An active learning algorithm could be the following : The machine questions the teacher about its response to a given input. It memorizes the answer and questions the teacher about whether the response has changed by reversing the state of one input unit, say unit i.The process is repeated until the response has changed indeed. The pupil considers that this unexpected change is surprising and that this state and the state I^{μ} it comes from are particularly interesting since the system must discriminate (give a different answer) for these close inputs. In actual fact they are interesting because if k is the label of input unit whose reversal makes the response of the teacher perceptron changes, the weight J_k is unambiguously determined. It is given by $J_k = \xi_k^{\mu} \xi^{\mu}$ where ξ_k^{μ} is the state of unit k in I^{μ} and ξ^{μ} the corresponding response of the teacher.

Obviously it is not permitted to go too far in asking questions to the environment. Would, for example, the pupil be allowed to directly question the teacher about its weights the problem of learning and generalizing would completely evaporate!

References

[1] Anshelevich, V., B. Amirikian, A. Lukashin. M. Frank-Kamenetskii, *On the ability of neural networks to perform generalization by induction.* Biol. Cybern., 125-128 (1989).

[2] Carnevali, P., S. Patarnello, *Exhaustive thermodynamical analysis of Boolean learning networks.* Europhys. Let. 4, 1199-1204 (1987).

[3] Changeux, J.P., T. Heidmann, P. Piette, *Learning by selection.* In "The Biology of Learning", Marler, P., H. Terrace (eds.), Springer, Berlin, 115-133 (1984).

[4] Gardner, E., *The space of interactions in neural network models.* J. Phys. A: Math. Gen. 21, 257-270 (1988).

[5] Gardner, E., B. Derrida, *Three unfinished works on the optimal storage capacity of networks.* J. Phys. A: Math. Gen. 22, 1983-1994 (1989).

[6] Gordon, M., P. Peretto, *The statistical distribution of Boolean gates in two-inputs, one-output multilayered neural networks.* J. Phys. A: Math. Gen. 23, 3061-3072 (1990).

[7] Grossman, T., R. Meir, E. Domany, *Learning by choice of internal representations.* Complex systems 2, 555, (1988).

[8] Hinton, G., T. Sejnowski, D. Ackley, *Boltzmann machines: Constraint satisfaction networks that learn.* Carnegie-Mellon University, Technical Report CMU-CS-84-119, Pittsburgh PA (1984).
[9] Judd, S., *On the complexity of loading shallow neural networks.* Journal of Complexity, 4, 177-192 (1988).
[10] Kohonen, T., *Self-organized formation of topologically correct feature maps.* Biol. Cybern. 43, 59-69 (1982).
[11] Krauth, W., M. Mézard, *Learning algorithms with optimal stability in neural networks.* J. Phys. A: Math. Gen. 20, L745-L751 (1987).
[12] Le Cun, Y., *A learning scheme for asymmetric threshold network.* In "Cognitiva 85" Cesta-AFCET (eds), Paris, 599-604 (1985).
[13] Levin, E., N. Tishby, S. Solla, *A statistical approach to learning and generalization in layered neural networks.* Proceeding of the IJCNN, Washington, IEEE, 2, 403 (1989).
[14] Mézard, M., J.P. Nadal, *Learning in feed-forward layered networks: The tiling algorithm.* J. Phys. A: Math. Gen. 22, 2191-2203 (1989).
[15] Mézard, M., S. Patarnello, *On the capacity of feed forward layered networks.* A preprint 89:24 from ENS, Paris, (1989).
[16] Peretto, P., *On the dynamics of memorization processes.* Neural Networks 1, 309-322 (1988).
[17] Peretto, P., *Learning learning sets in neural networks.* Int. J. Neural Syst. 1, 31-41 (1989).
[18] Rumelhart, D., T. Hinton, R. Williams, *Learning representations by back-propagating errors.* Nature, 323, 533-536 (1986).
[19] Sompolinsky, H., N. Tishby, H. Seung, *Learning from examples in large neural networks.* Phys. Rev. Lett. 65, 1683-1686 (1990).
[20] Vallet, F., *The Hebb rule for learning linearly separable Boolean functions: Learning and generalization.* Europhys. Lett. 8, 747-751 (1989).

ON THE R.E.M. AND THE G.R.E.M.

P. PICCO
C.N.R.S. C.P.T.
Luminy case 907
13288 Marseille
France

Introduction

Random Energy model and generalized Random Energy model was introduced by B. Derrida [1], [2], [3], [4] as simple and solvable models for spin glasses.

Spin Glasses are disordered magnetic systems and a mean field description of such system is given by the Sherrington and Kirkpatrick model [5]. This model is defined on a one dimensional lattice for Ising spin $\sigma = \pm 1$ by the following Hamiltonian:

$$H(\underline{\sigma}) = - \sum_{1 \leq i < j \leq N} \frac{J_{ij}}{\sqrt{N}} \sigma_i \sigma_j$$

where the J_{ij} are independent Gaussian random variables with zero mean and variance 1.

Concerning this model we refer the reader to the book of Mezard-Parisi-Virasoro [6] where she (he) can find complete references.

Let us remark that $\mathbb{E}(H(\sigma)) = 0$ and

$$\begin{aligned} \mathbb{E}(H(\underline{\sigma})H(\underline{\sigma}')) &= \frac{1}{N} \sum_{1 \leq i < j \leq N} \sum_{1 \leq k < \ell \leq N} \sigma_i \sigma_j \sigma'_k \sigma'_\ell \mathbb{E}(J_{ij} J_{k\ell}) \\ &= \frac{1}{N} \sum_{1 \leq i < j \leq N} \sigma_i \sigma'_i \sigma_j \sigma'_j \\ &= \frac{1}{N} [\frac{1}{2} (\sum_{i=1}^{N} \sigma_i \sigma'_i)^2 - \frac{N}{2}] \end{aligned}$$

If we set

$$q_{\sigma\sigma'} = \frac{1}{N} \sum_{i=1}^{N} \sigma_i \sigma'_i$$

E. Goles and S. Martínez (eds.), Statistical Physics, Automata Networks and Dynamical Systems, 173–207.

we get

$$\mathbb{E}(H(\underline{\sigma})H(\underline{\sigma}')) = \frac{1}{2N}((Nq_{\sigma\sigma'})^2 - N)$$
$$= \frac{N}{2}(q_{\sigma\sigma'}^2 - \frac{1}{N})$$

in particular $\mathbb{E}((H(\sigma))^2) = \frac{N-1}{2}$

Let us define the partition function by

$$Z_N = \sum_{\{\sigma\}} e^{-\beta H(\sigma)}$$

If we want to consider normalized Gaussian random variables we write

$$Z_N = \sum_{\{\sigma\}} e^{\beta\sqrt{\frac{N-1}{2}}(E_\sigma)}$$

where

$$E_\sigma = \sqrt{2/(N-1)N} \sum_{1 \leq 1 < j \leq N} J_{ij}\sigma_i\sigma_j$$

In which case

$$\mathbb{E}(E_\sigma E_{\sigma'}) = \frac{N}{N-1}[q_{\sigma\sigma'}^2 - \frac{1}{N}]$$
$$= \frac{Nq_{\sigma\sigma'}^2 - 1}{N-1}$$

and $\mathbb{E}(E_\sigma^2) = 1$.

We remark that in the general case $\mathbb{E}(E_\sigma E'_\sigma) \neq 0$ that is the random variables $\{E_\sigma\}_\sigma$ case correlated.

The REM and GREM are simplification of this model, in the first one we assume that the random variables E_σ are independent. This assumption allows us to solve the model who reveal a very rich structure.

In the second case (the GREM) we allow the random variables to be correlated in a particular way, the structure of the model becomes richer. In particular the correlation depends on $n-1$ parameters, since the possible values of the correlation in the S.K. model are $[\frac{N}{2}] + 1$, one can hope to recover the S.K. model by taking n of order N, the main problem is that the proof we will gave depends on the fact that n is a most of order $\frac{N}{(\log N)^{1+\epsilon}}$. The $(\log N)^{1+\epsilon}$ is for fixing the ideas, it could be improved, but in any case our proof depends on the fact that $\frac{n}{N} \to 0$ when $N \to \infty$. In the case $\frac{n}{N} \to$ cte $\neq 0$, fluctuations that come from the number of hierarchies (the factor n) destroy the picture we will gave later.

1. Definition of the Models

The REM is defined by the following partition function

$$Z_N(\beta) = \sum_{i=1}^{2^N} \exp \beta\sqrt{N} X_i \tag{1.1}$$

where X_i are independent normalized Gaussian random variables. That is

$$I\!P(X_i \geq x) = \frac{1}{\sqrt{2\pi}} \int_x^{+\infty} e^{-t^2/2} dt \tag{1.2}$$

and for any family of bounded real functions $(f_i)_{i=1}^{2^N}$

$$I\!E\left(\prod_{i=1}^{2^N} f_i(x_i)\right) = \prod_{i=1}^{2^N} I\!E(f_i(x_i)) \tag{1.3}$$

We shall define the finite volume free energy:

$$F_N(\beta) = \frac{1}{N} \log Z_N(\beta) \tag{1.4}$$

The GREM is defined in the following way: for any integers n and N let $a_i \geq 0$ and $\alpha_i \geq 1$ $i = 1, ..., n$ be real positive numbers such that

$$\sum_{i=1}^{n} a_i = 1, \quad \sum_{i=1}^{n} \log \alpha_i = \log 2 \tag{1.5}$$

Let $(\Omega, \Sigma, I\!P)$ be a probability space such that for any $n \in I\!N$ and $N \in I\!N$ then exists a family of $\alpha_1^N + \alpha_1^N \alpha_2^N + \alpha_1^N \alpha_2^N \alpha_3^N + ... + \alpha_1^N \cdots \alpha_n^N$ independent, normalized Gaussian random variables

$$\varepsilon^j_{k_1,...,k_j} \in \mathcal{N}(0,1) \text{ for } j = 1, ..., n, \quad k_j \in 1, ..., \alpha_j^N$$

The GREM is defined by the partition function

$$Z_{n,N}(\beta) = \sum_{k_1=1}^{\alpha_1^N} ... \sum_{k_n=1}^{\alpha_n^N} \exp \beta\sqrt{N}\left(\sum_{j=1}^{n} \sqrt{a_j} \varepsilon^j_{k_1,...,k_j}\right) \tag{1.6}$$

Let us remark that $\sum_{i=1}^{2^N}$ in the REM and in the GREM $\sum_{\alpha_1^N} \dots \sum_{\alpha_n^N}$ correspond to the 2^N spin configurations of an Ising spin (that is $\sigma_i = \pm 1$) model with N sites.

We shall define for the GREM the finite volume free energy by

$$F_{n,N}(\beta) = \frac{1}{N} \log Z_{n,N}(\beta) \tag{1.7}$$

We shall also define the following subset of $\mathbb{R}^k$ (for any $k = 1, 2, ..., n$)

$$\tilde{\xi}(k) = \{(x_i),\ i = 1, ..., k \quad \sum_{i=1}^{j} \frac{X_i^2}{2} \leq \sum_{i=1}^{j} \log \alpha_i \quad \forall j \in 1, ..., k\} \tag{1.8}$$

and $\|x\|^2 = \sum_{i=1}^{n} x_i^2$ is the Euclidean norm of $\mathbb{R}^n$. Our first result is [7] [8].

Theorem 1.1. Let m^* be the n-dimensional vector $m^* \equiv (m_i^*)_{i=1}^n = (\beta\sqrt{a_i})_{i=1}^n$ and let m be the point of the compact convex subset of $\mathbb{R}^n$: $\tilde{\xi}(n)$ at minimal distance from m^*.

Then for any $\beta > 0$

$$F_n \equiv \lim_{N\to\infty} \frac{1}{N} \log Z_{n,N}(\beta) = \frac{\|m^*\|^2}{2} - \frac{\|m - m^*\|^2}{2} + \ell n 2 \tag{1.9}$$

almost surely and in $\mathbb{L}^p(\Omega, \Sigma, \mathbb{P})$ for any $1 \leq p < +\infty$.

We shall recall some basic tools on convergence of random variables.

Let (X_n) be an arbitrary family of real random variables an a probability space $(\Omega, \Sigma, \mathbb{P})$. The following convergences can occur.

I. $X_n \to X$ in $\mathbb{P}$ Probability (or $\mathbb{P}$ Stochastly or in measure) if $\forall \varepsilon > 0$

$$\lim_{n\to\infty} \mathbb{P}(|X_n - X| > \varepsilon) = 0$$

II. $X_n \to X$ $\mathbb{P}$ almost everywhere (or $\mathbb{P}$ almost surely) if

$$\mathbb{P}(\{\lim_{n\to\infty} X_n = X\}) = 1$$

(which is equivalent to $\mathbb{P}(\{\lim_{n\to\infty} X_n \neq X\} \cup \{\lim X_n \text{ does not exists }\} = 0)$

III. $X_n \to X$ in $\mathbb{L}^p(\Omega, \Sigma, \mathbb{P})$

$$\mathbb{E}(|X_n - X|^p)^{\frac{1}{p}} \to 0$$

It is standard result that II $\Rightarrow$ I and III $\Rightarrow$ I (see any book on measure theory, probability theory); in general I $\not\Rightarrow$ III as the following example shows:

Let

$$X_n = \begin{cases} 0 & \text{with probability } 1 - \frac{1}{n^\alpha} \\ n^\beta & \text{with probability } \frac{1}{n^\alpha} \end{cases} \tag{1.10}$$

then $X_n \to 0$. $I\!P$ a.e. and in $I\!P$ Probability.

But

$$I\!E(X_n^p) = \frac{n^{p\beta}}{n^\alpha} = n^{p\beta-\alpha} \tag{1.11}$$

if $p\beta - \alpha < 0$ then $X_n \to 0$ in $I\!L^p$
if $p\beta - \alpha \geq 0$ X_n does not converge in $I\!L^p$.

Then exists a criterium:

If (X_n) is a family of uniform $I\!L^p$ integrable real random variables (that is $\lim_{\alpha\to\infty} \sup_{n\geq n_0} \int_{|x_k|^p \geq \alpha} |Y_n|^p dI\!P = 0$) and $X_n \to X$ in $I\!P$ probability then $X_n \to X$ in $I\!L^p$.

The fact that I $\not\Rightarrow$ II is less trivial. The following counterexample is fundamental

Let $\Omega = [0,1]$, $I\!P$ the Lebesgue measure and Σ the Lebesgue σ-algebra.

Let us define for $1 \leq k \leq 2^N$

$$\begin{aligned} f_k^N(n) &= 1\!\!1_{[\frac{k-1}{2^N}, \frac{k}{2^N}]}(x) \\ &= \begin{cases} 1 & \text{if } x \in [\frac{k-1}{2^N}, \frac{k}{2^N}] \\ 0 & \text{otherwise} \end{cases} \end{aligned} \tag{1.12}$$

If $p \in 2^{\ell+1} - 1$, $2^{\ell+2} - 1$ for some $\ell \in I\!N$ then p can be written $2^{\ell+1} - 1 + k(p)$ for some $k(p) \in 1, ..., 2^{\ell+1}$; we define

$$g_p(x) = f_k^{\ell+1}(x). \tag{1.13}$$

(The reader is invited to draw the graph of this function)

Since, if $p \in 2^{\ell+1}$, $2^{\ell+2} - 1$ and ε is strictly positive it is easy to check that

$$I\!P(g_p > \varepsilon) = \frac{1}{2^{\ell+1}} \tag{1.14}$$

and if $p \to \infty$ then $\ell \to \infty$, we get

$$\lim_{p\to\infty} I\!P(g_p > \varepsilon) = 0. \tag{1.15}$$

Since $\|g_p\|_{\ell_1} = \frac{1}{2^{\ell+1/\ell_1}} \to 0 \quad \forall 1 \leq \ell_1 < \infty$ we get: $g_p \to 0$ in all $I\!L^p$ $1 \leq p < \infty$.

We will show the following results:

$$1. \quad \mathbb{P}(\liminf g_p = 0) = 1$$

$$2. \quad \mathbb{P}(\limsup g_p = 1) = 1$$

where

$$\liminf g_p = \lim_{k\to\infty} \inf_{p\geq k} g_p = \sup_{n\geq 1} \inf_{k\geq n} g_k$$

and

$$\limsup g_p = \lim_{k\to\infty} \sup_{p\geq k} g_p = \inf_{n\geq 1} \sup_{k\geq n} g_k$$

Let us first prove 1.

Since $g_p \geq 0$ we get $\liminf g_p \geq 0$; on the other hand we have $\lim_{p\to\infty} \mathbb{P}(g_p = 0) = 1$ we will prove that $\mathbb{P}(\{g_p = 0\} \text{ i.o}) = 1$ where $\{g_p = 0\} \text{ i.o.} = \bigcap_{n\geq 1} \bigcup_{k=n}^{\infty} \{g_k = 0\}$. Since $B_n = \bigcup_{k=n}^{\infty} \{g_k = 0\}$ is decreasing with n, we get, by σ additivity,

$$\mathbb{P}(\bigcap_n B_n) = \lim_{n\to\infty} \mathbb{P}(B_n)$$

$$= \lim_{n\to\infty} \mathbb{P}(\bigcup_{k=n}^{\infty} \{g_k = 0\}) \geq \lim_{n\to\infty} \mathbb{P}(\{g_n = 0\}) = 1.$$

Let us prove $\mathbb{P}(\limsup g_p = 1) = 1$ that is $\mathbb{P}(\bigcap_{k=1}^{\infty} \bigcup_{p=k}^{\infty} \{g_p = 1\}) = 1$. Since for any ℓ

$$\bigcup_{p=2^{\ell}+1}^{2^{\ell+2}-1} \{g_p = 1\} = [0,1]$$

and given a k there exists $\ell = \ell(k)$ increasing with k such that $k \leq 2^{p(k)+1}$ (take $\ell(k) = \log_2 k$) we get

$$\mathbb{P}(\bigcup_{p=k}^{\infty} \{g_p = 1\}) \geq \mathbb{P}(\bigcup_{p=2^{\ell(k)}+1}^{2^{\ell(k)+2}} \{g_p = 1\}) = 1$$

therefore

$$\lim_{k\to\infty} \mathbb{P}(\bigcup_{p=K}^{\infty} \{g_p = 1\}) = 1.$$

This last example looks like a measure theoretical ones; in the REM the following is true [9]

Theorem 1.2. If $\beta c = \sqrt{2\log 2}$ and

$$\mathcal{H}_N(\beta) = \frac{\beta_c}{\beta} \frac{\log[Z_N e^{-NF(\beta)}]}{\log N} \tag{1.16}$$

(this is, in some sense, the finite size corrections of order $\frac{\log N}{N}$ to the free energy). Then, if $0 < \beta < \beta_c$

$$\lim \mathcal{H}_N(\beta) = 0 \quad I\!\!P \text{ almost everywhere}$$

if $\beta \geq \beta_c$

$$\limsup \mathcal{H}_N(\beta) = \frac{1}{2} \quad I\!\!P \text{ almost everywhere}$$

$$\liminf \mathcal{H}_N(\beta) = -\frac{1}{2} \quad I\!\!P \text{ almost everywhere.}$$

It can be proved that $\lim \mathcal{H}_N(\beta) = -\frac{1}{2}$ in Probability [4].

The last result for the REM concern the "renormalized random variables"

$$\xi_i^N = \exp[\beta\sqrt{N}X_i - \beta\beta_c N + \frac{1}{2}\frac{\beta}{\beta_c}N] \tag{1.17}$$

for $i = 1, ..., 2^N$ which converge in law to a Poisson Point Process that is:

$$\mathcal{N}_t^{(N)} = \sum_{i=1}^{2^N} 1\!\!1_{\geq g_N(t)}(\xi_i^N) \tag{1.18}$$

converges in law to the counting measure of a Poisson Point Process, if $g_N(t) = \exp \frac{\beta t}{\beta} - \frac{\beta}{\beta_c} \log \beta_c \sqrt{2\pi}$, this P.P.P. have intensity exp $(-t)$.

Similar result exists for the GREM we refer to the original paper [9] and to the paper of Ruelle [10] for details and applications of this result. We will not give the proof of these last results. See [7] and [9].

We will prove the Theorem 1.3.

Let $\beta_c = \sqrt{2\log 2}$ and

$$F(\beta) = \begin{cases} \frac{\beta'}{2} + \frac{\beta_c^2}{2} & 0 < \beta \leq \beta_c \\ \beta\beta_c & \beta \geq \beta_c \end{cases} \tag{1.19}$$

then $\forall\beta > 0 \lim F_N(\beta) = F(\beta)$ $I\!P$ almost everywhere and in $I\!L^p(\Omega, \Sigma, I\!P)$ $\forall 1 \le p < \infty$.

The first problem is the following.

Let us consider

$$Z_N = \sum_{i=1}^{2^N} e^{\beta\sqrt{N}X_i}$$

If the family $(X_i)_{i=1}^{2^N}$ have fluctuations higger than $\sqrt{N}$, say $(N)^{\frac{1}{2}+\varepsilon}$ then F does not exist.

It is natural to consider the random variable

$$\xi = \max_{i=1,\dots,2^N} |X_i| \tag{1.20}$$

because

$$Z_N \le 2^N \exp\beta\sqrt{N}\xi$$

and

$$Z_N \ge \exp\beta\sqrt{N}\xi$$

Let us estimate

$$\begin{aligned} I\!P(\max_{i=1,\dots,2^N} |X_i| \ge \gamma\sqrt{N}) &= I\!P(\max_{i=1,\dots,2^N} \frac{X_i^2}{2} \ge \frac{N\gamma^2}{2}) \\ &\le e^{-t\frac{N\gamma^2}{2}} I\!E(e^{t \max_{i=1,\dots,2^N} \frac{X_i^2}{2}}). \end{aligned} \tag{1.21}$$

Since

$$I\!E(e^{t \max_{i=1,\dots,2^N} \frac{X_i^2}{2}}) = I\!E(\max_{i=1,\dots,2^N} e^{t\frac{X_i^2}{2}}) \le I\!E(\sum_{i=1}^{2^N} e^{\frac{tX_i^2}{2}}) = 2^N I\!E(e^{\frac{tX^2}{2}}). \tag{1.22}$$

we get

$$I\!P(\max_{i=1,\dots,2^N} |X_i| \ge \gamma\sqrt{N}) \le 2^N e^{-\frac{N\gamma^2}{2}t} \frac{1}{\sqrt{1-t}} \tag{1.23}$$

This probablity goes to zero if $\frac{\gamma^2}{2} = (1+2\varepsilon)\log 2$ and $t = 1-\varepsilon$ with $0 < \varepsilon < \frac{1}{2}$; therefore we get

$$2^N e^{-\frac{N\gamma^2}{2}t} \frac{1}{\sqrt{1-t}} = \frac{1}{\sqrt{\varepsilon}} e^{-N(\log 2)(\varepsilon - 2\varepsilon^2)} \tag{1.24}$$

Now for a fixed $\frac{1}{2} > \varepsilon > 0$ if $N \to \infty$

$$I\!P(\max |X_i| \ge (1+2\varepsilon)^{\frac{1}{2}} \sqrt{N 2 \log 2}) \le \frac{1}{\sqrt{\varepsilon}} e^{-N \log 2(\varepsilon - 2\varepsilon')} \tag{1.25}$$

which goes to zero in a summable way.

At this point we use the Borel Cantelli lemma

Lemma. If A_n is a sequence of events and $\sum_n \mathbb{P}(A_n) < +\infty$ then

$$\mathbb{P}(A_n \text{ i.o }) = \mathbb{P}(\bigcap_m \bigcup_{k=m}^{\infty} A_k) = 0.$$

Remark. This means also

$$\mathbb{P}(\bigcup_m \bigcap_{k=m}^{\infty} A_k^c) = 1$$

that is there exist a subset Ω_0 such that $\mathbb{P}(\Omega_0) = 0$ and for any $\omega \in \Omega \setminus \Omega_0$ then exists a finite $m = m(\omega)$ such that for any $k \geq m$

$$A_k^c \text{ occurs}$$

Proof.

$$\mathbb{P}(A_n \text{ i.o. }) \leq \mathbb{P}(\bigcup_{k=m}^{\infty} A_k) \leq \sum_{k=m}^{\infty} \mathbb{P}(A_k)$$

therefore

$$0 \leq \mathbb{P}(A_n \text{ i.o }) \leq \lim_{m\to\infty} \sum_{k=m}^{\infty} \mathbb{P}(A_k) = 0 \quad \blacksquare$$

Using this lemma we can deduce for almost every ω there exist a $n = n(\omega)$ (finite) such that for any $N \geq n(\omega)$

$$0 \leq \max_{i=1,\ldots,2^N} |X_i| \leq \sqrt{2N(1+\varepsilon)\log 2}.$$

Be careful, the $n = n(\omega)$ depends on ω and ε.

Let us remark that we have not used the fact that the random variables X_i are independent.

We will first prove that

$$F_N(\beta) \leq F(\beta) + \varepsilon \quad \mathbb{P} \text{ a.e} \tag{1.26}$$

that is for almost all ω there exist a finite $n = n(\omega)$ such that if $N \geq n(\omega)$ the previous inequality is true.

Let us write

$$\begin{aligned} Z_N = Z_N \mathbb{1}_{\{\forall i \in 1,2,\ldots,2^N,\ |X_i| \leq \sqrt{2N(\log 2)(1+\varepsilon)}\}} \\ + Z_N \mathbb{1}_{\{\exists i \in 1,\ldots,2^N, |X_i| \geq \sqrt{2N(\log 2)(1+\varepsilon)}\}} \end{aligned} \tag{1.27}$$

then

$$\begin{aligned} \mathbb{P}(Z_N \mathbb{1}_{\{\exists i\}} > \frac{\varepsilon}{2}) = \mathbb{P}(\{Z_N \mathbb{1}_{\{\exists i\}} \geq \frac{\varepsilon}{2}\} \cap \{\mathbb{1}_{\{\exists i\}} = 1\}) \\ + \mathbb{P}(\{Z_N \mathbb{1}_{\{\exists i\}} \geq \frac{\varepsilon}{2}\} \cap \{\mathbb{1}_{\{\exists i\}} = 0\}) \end{aligned} \tag{1.28}$$

since $\frac{\varepsilon}{2} > 0$ and $\{Z_N \mathbb{1}_{\{\exists i\}} \geq \frac{\varepsilon}{2}\} \cap \{\mathbb{1}_{\{\exists i\}} = 0\} = \phi$. We get

$$\begin{aligned} \mathbb{P}(Z_N \mathbb{1}_{\{\exists i\}} > \frac{\varepsilon}{2}) &\leq \mathbb{P}(\mathbb{1}_{\{\exists i\}} = 1) \\ &= \mathbb{P}(\max |X_i| \geq \sqrt{2N(\log 2)(1+\varepsilon)} \\ &\leq \frac{1}{\sqrt{\frac{\varepsilon}{2}}} \exp - N(\log 2)(\frac{\varepsilon}{2} - 2(\frac{\varepsilon}{2})^2) \end{aligned} \tag{1.29}$$

On the other hand

$$\mathbb{P}(Z_N \mathbb{1}_{\{\forall i\}} \geq e^{\gamma N} \mathbb{E}(Z_N \mathbb{1}_{\{\forall i\}})) \leq e^{-\gamma N} \tag{1.30}$$

which is summable, therefore using the Borel Cantelli lemma we get if $N \geq N(\omega)$

$$Z_N \leq \varepsilon + e^{\gamma N} \mathbb{E}(Z_n \mathbb{1}_{\{\forall i\}}) \quad \mathbb{P} \text{ a.e.} \tag{1.31}$$

Let us compute

$$\begin{aligned} &\mathbb{E}((\sum_{i=1}^{2^N} e^{\beta\sqrt{N} X_i}) \mathbb{1}_{\{\forall i, |X_i| \leq \sqrt{2N(\log 2)(1+\varepsilon)}\}}) \\ &= \sum_{i=1}^{2^N} \mathbb{E}(e^{\beta\sqrt{N} X_i} \mathbb{1}_{\{\forall i, |X_i| \leq \sqrt{2N(\log 2)(1+\varepsilon)}\}}) \\ &\leq 2^N \mathbb{E}(e^{\beta\sqrt{N} X} \mathbb{1}_{\{|X| \leq \sqrt{2N(\log 2)(1+\varepsilon)}\}}) \end{aligned} \tag{1.32}$$

That is

$$2^N \int_{-\sqrt{(2N)(1+\varepsilon)\log 2}}^{+\sqrt{2N(1+\varepsilon)(\log 2)}} e^{\beta\sqrt{N}X-\frac{x^2}{2}} \frac{dx}{\sqrt{2\pi}}$$
$$= e^{(\frac{\beta_c^2}{2}+\frac{\beta^2}{2})N} \int_{-\sqrt{(2N)(1+\varepsilon)\log 2}}^{\sqrt{(2N)(1+\varepsilon)\log 2}} e^{-\frac{1}{2}(x-\beta\sqrt{N})^2} \frac{dx}{\sqrt{2\pi}} \tag{1.33}$$

Calling $\beta_c(\varepsilon) = \sqrt{2(1+\varepsilon)\log 2)}$ we have to estimate

$$\int_{-\beta_c(\varepsilon)\sqrt{N}}^{\beta_c\sqrt{N}} e^{-\frac{1}{2}(x-\beta\sqrt{N})^2} dx \tag{1.34}$$

Let us remark that if $\beta < \beta_c$ the center of the Gaussian distribution is inside the integration interval, if $\beta > \beta_c$ the center of the Gaussian is outside this interval. Therefore we can expect that if $\beta < \beta_c$ the previous integral goes to one when $N \to \infty$ and if $\beta > \beta_c$ the previous integral goes to zero.

If $\beta < \beta_c$ we write

$$\int_{-\beta_c(\varepsilon)\sqrt{N}}^{\beta_c\sqrt{N}} e^{-\frac{1}{2}(x-\beta\sqrt{N})^2} \frac{dx}{\sqrt{2\pi}} = \int_{-(\beta+\beta_c(\varepsilon))\sqrt{N}}^{(\beta_c-\beta)\sqrt{N}} e^{-\frac{1}{2}y^2} \frac{dy}{\sqrt{2\pi}}$$
$$= 1 - \int_{(\beta_c-\beta)\sqrt{N}}^{\infty} e^{-\frac{x}{2}} \frac{dx}{\sqrt{2\pi}} - \int_{(\beta_c(\varepsilon)+\beta)\sqrt{N}}^{\infty} e^{-\frac{x}{2}} \frac{dx}{\sqrt{2\pi}} \tag{1.35}$$

We will use the following estimate: $\forall \alpha \geq 0$

$$\Phi(\alpha) \equiv \int_{\alpha}^{\infty} e^{-\frac{x^2}{2}} \frac{dx}{\sqrt{2\pi}} \geq \frac{e^{-\frac{1}{2}\alpha^2}}{(\alpha+1)\sqrt{2\pi}} \equiv g(\alpha) \tag{1.36}$$

Let us prove this inequality:
We remark that $g(\infty) = 0$ and $\Phi(\infty) = 0$. In the other hand

$$g'(t) = \frac{1}{\sqrt{2\pi}}\left(\frac{1}{(1+t)^2} - \frac{t}{(1+t)}\right)e^{-\frac{t^2}{2}} \geq -\frac{t}{1+t}\frac{1}{\sqrt{2\pi}}e^{-\frac{t^2}{2}} \tag{1.37}$$

Since $\frac{t}{1+t} \leq 1 \ \forall t \geq 0 \quad g'(t) \geq -\frac{1}{\sqrt{2\pi}} e^{-\frac{1}{2}+2} = \phi'(t)$

Therefore since $\phi(\infty) = g(\infty) = 0$ and $0 > g'(t) > \phi'(t)$ we get $\phi(t) \geq g(t)$.

Let us prove an other inequality

$$\int_{\alpha}^{\infty} e^{-\frac{x^2}{2}} \frac{dx}{\sqrt{2\pi}} \leq \frac{1}{(\sqrt{2\pi})\alpha} e^{-\frac{\alpha^2}{2}} \tag{1.38}$$

We have

$$\int_{\alpha}^{\infty} e^{-\frac{x^2}{2}} \frac{dx}{\sqrt{2\pi}} = \int_{\alpha}^{\infty} \frac{x}{x} e^{-\frac{x^2}{2}} \frac{dx}{\sqrt{2\pi}} \leq \frac{1}{\alpha} \int_{\alpha}^{\infty} x e^{-\frac{x^2}{2}} \frac{dx}{\sqrt{2\pi}} = \frac{e^{\frac{-\alpha^2}{2}}}{\sqrt{2\pi}\alpha} \tag{1.39}$$

Therefore if $\beta < \beta_c$

$$\begin{aligned} &\int_{-\beta_c(\varepsilon)\sqrt{N}}^{\beta_c(\varepsilon)\sqrt{N}} e^{-\frac{1}{2}(x-\beta\sqrt{N})^2} \frac{dx}{\sqrt{2\pi}} \\ &\leq 1 - \frac{e^{-\frac{N}{2}(\beta_c-\beta)^2}}{(1+(\beta_c-\beta)\sqrt{N})\sqrt{2\pi}} - \frac{e^{-\frac{N}{2}(\beta_c-\beta)^2}}{(1+(\beta_c-\beta)\sqrt{N})\sqrt{2\pi}} \leq 1 \end{aligned} \tag{1.40}$$

In the case $\beta \geq \beta_c$, the center of the gaussian is outside the interval of integration, the main contribution of the integral comes from the right extreme of this interval. This can be seen by the following arguments

$$\begin{aligned} (x-\beta\sqrt{N})^2 &= (x-\beta_c\sqrt{N}-(\beta\sqrt{N}-\beta_c\sqrt{N}))^2 \\ &= (x-\beta_c\sqrt{N})^2 + (\beta-\beta_c)^2 N - 2(X-\beta_c\sqrt{N})(\beta-\beta_c)\sqrt{N} \end{aligned} \tag{1.41}$$

Since $x < \beta_c\sqrt{N}$, $\beta \geq \beta_c$ we get

$$-2(x-\beta_c\sqrt{N})(\beta-\beta_c) \geq 0$$

this implies

$$(x-\beta\sqrt{N})^2 \geq (x-\beta_c\sqrt{N})^2 + (\beta-\beta_c)^2 N \tag{1.42}$$

Therefore

$$\int_{-\beta_c\sqrt{N}}^{\beta_c\sqrt{N}} e^{-\frac{1}{2}(x-\beta\sqrt{N})^2} \frac{dx}{\sqrt{2\pi}} \leq e^{-\frac{1}{2}(\beta-\beta_c)^2 N} \int_{-\beta_c\sqrt{N}}^{\beta_c\sqrt{N}} e^{-(x-\beta_c\sqrt{N})^2} \frac{dx}{\sqrt{2\pi}} \tag{1.43}$$

Now

$$\int_{-\beta_c\sqrt{N}}^{\beta_c\sqrt{N}} e^{-\frac{1}{2}(x-\beta_c\sqrt{N})^2} \frac{dx}{\sqrt{2\pi}} = \int_0^{2\beta_c\sqrt{N}} e^{-\frac{t^2}{2}} \frac{dt}{\sqrt{2\pi}}$$
$$= \frac{1}{2} - \int_{2\beta_c\sqrt{N}}^{\infty} e^{-\frac{t^2}{2}} \frac{dt}{\sqrt{2\pi}} \leq \frac{1}{2} \tag{1.44}$$

and

$$\geq \frac{1}{2} - \frac{e^{-4\frac{\beta_c^2 N}{2}}}{(2\beta_c\sqrt{N})\sqrt{2\pi}} \tag{1.45}$$

Collecting 1.31, 1.32, 1.40 and 1.44 we get

$$\begin{aligned} e^{\gamma N}\mathbb{E}\{Z_N \mathbb{1}_{\{\forall i\}}\} &\leq e^{\gamma N} e^{\frac{\beta_c^2}{2}N+\frac{\beta^2}{2}N} && \text{if } \beta \leq \beta_c \\ &\leq e^{\gamma N} e^{\frac{\beta_c^2}{2}N+\frac{\beta^2}{2}N} e^{-\frac{1}{2}(\beta-\beta_c)^2 N} && \text{if } \beta > \beta_c \\ &= e^{\gamma N} e^{+\beta\beta_c N} \end{aligned} \tag{1.46}$$

By taking the logarithm and dividing by N we get 1.26. We will prove that for almost all ω $\exists N(\omega)$ such that $N \geq N(\omega)$ implies

$$F_N(\beta) \geq F(\beta) - \varepsilon.$$

The idea, is to restrict the range of the random variables X_i to be in an interval, in such a way that we get the dominant contribution to Z_N, in fact to $\mathbb{E}(Z_N \mathbb{1}_{\{\forall i\}})$. By the previous computations the main contribution comes from $X \simeq \beta\sqrt{N}$ if $\beta \leq \beta_c$ and from $X \simeq \beta_c\sqrt{N}$ if $\beta \geq \beta_c$.

Let us define

$$\Delta(\beta) = \begin{cases} [\beta - 2\varepsilon, \beta - \varepsilon]\sqrt{N} & \text{if } \beta \leq \beta_c \\ [\beta_c - 2\varepsilon, \beta_c - \varepsilon]\sqrt{N} & \text{if } \beta > \beta_c \end{cases} \tag{1.47}$$

and call $\tilde{\beta} = \{ \begin{array}{l} \beta \text{ if } \beta \leq \beta_c \\ \beta_c \text{ if } \beta > \beta_c \end{array}$

It is easy to check that

$$Z_N \geq \sum_{i=1}^{2^N} \mathbb{1}_{\Delta(\beta)}(x_i) e^{\beta\sqrt{N}X_i} \geq e^{\beta\sqrt{N}(\tilde{\beta}-2\varepsilon)\sqrt{N}} \sum_{i=1}^{2^N} \mathbb{1}_{\Delta(\beta)}(x_i) \tag{1.48}$$

Let us prove that, for almost all ω if $N \geq N(\omega)$ then

$$\sum_{i=1}^{2^N} \mathbb{1}_{\Delta(\beta)}(x_i) \geq \eta 2^N \mathbb{E}(\mathbb{1}_{\Delta(\beta)}(x_i))$$

if $0 < \eta < 1$.

Since

$$\begin{aligned}
&\{\sum \mathbb{1} \leq \eta 2^N \mathbb{E}(\mathbb{1})\} = \{\sum(\mathbb{1} - \mathbb{E}(\mathbb{1})) \leq (\eta - 1)2^N \mathbb{E}(\mathbb{1})\} \\
&\subset \{\sum(\mathbb{1} - \mathbb{E}(\mathbb{1})) \leq -(1-\eta)2^N \mathbb{E}(\mathbb{1})\} \cup \{\sum(\mathbb{1} - \mathbb{E}(\mathbb{1})) \leq (1-\eta)2^N \mathbb{E}(\mathbb{1})\} \\
&= \{|\sum(\mathbb{1} - \mathbb{E}(\mathbb{1}))| \geq (1-\eta)2^N \mathbb{E}(\mathbb{1})\}
\end{aligned} \tag{1.49}$$

We get

$$\mathbb{P}(\sum \mathbb{1} \leq \eta 2^N \mathbb{E}(\mathbb{1})) \leq \mathbb{P}(\sum(\mathbb{1} - \mathbb{E}(u))^2 \geq (1-\eta)^2 2^{2N}(\mathbb{E}(\mathbb{1}))^2) \tag{1.50}$$

$$\leq \frac{\mathbb{E}((\sum(\mathbb{1} - \mathbb{E}(\mathbb{1})))^2)}{(1-\eta)^2 2^{2N}(\mathbb{E}(\mathbb{1}))^2}$$

by Markov inequality.

The righthand side of 1.50 is nothing but

$$\frac{2^N(\mathbb{E}(\mathbb{1}))(1 - \mathbb{E}(\mathbb{1}))}{(1-\eta)^2 2^{2N}(\mathbb{E}(\mathbb{1}))^2} \leq \frac{1}{(1-\eta)^2 2^N \mathbb{E}(\mathbb{1})} \tag{1.51}$$

Let us estimate

$$\begin{aligned}
2^N \mathbb{E}(\mathbb{1}) = 2^N \int_{(\tilde{\beta}-2\varepsilon)\sqrt{N}}^{(\tilde{\beta}-\varepsilon)\sqrt{N}} e^{-\frac{x^2}{2}} \frac{dx}{\sqrt{2\pi}} \geq \frac{2^N \varepsilon \sqrt{N}}{\sqrt{2\pi}} e^{-\frac{|(\tilde{\beta}-\varepsilon)\sqrt{N}|^2}{2}} \\
= \varepsilon\sqrt{N} e^{(\frac{\beta_c^2}{2} - \frac{\tilde{\beta}^2}{2})N} e^{\varepsilon\tilde{\beta}N - \frac{\varepsilon^2 N}{2}}
\end{aligned} \tag{1.52}$$

if $\varepsilon < 2\tilde{\beta}$ then $\varepsilon\tilde{\beta} - \frac{\varepsilon^2}{2} = \varepsilon(\tilde{\beta} - \frac{\varepsilon}{2}) > 0$ and

$$\forall \beta > 0 \quad 2^N \mathbb{E}(\mathbb{1}) > e^{N(\varepsilon)(\tilde{\beta} - \frac{\varepsilon}{2})} \quad \to \infty \tag{1.53}$$

and therefore

$$\frac{1}{2^N \mathbb{E}(\mathbb{1})} \leq \frac{e^{-N(\varepsilon)(\tilde{\beta} - \frac{\varepsilon}{2})}}{\varepsilon\sqrt{N}}$$

which is the general term of a sumable series.

We conclude by applying the Borel Cantelli lemma.

Now we will prove the convergence in all $\mathbb{L}^P(\Omega, \Sigma, \mathbb{P})$ for $1 \leq p < \infty$. By using the uniform $\mathbb{L}^p$ integrability criterium, it is sufficient to prove

$$\lim_{\alpha\to\infty} \sup_{N\geq n} \int_{|F_N(\beta)|^p>\alpha} |F_N(\beta)|^P d\mathbb{P} = 0 \tag{1.54}$$

Since

$$F_N(\beta) \leq \log 2 + \frac{1}{\sqrt{N}}\beta \max_{i=1,\dots,2^N} X_i \tag{1.55}$$

and

$$F_N(\beta) \geq \frac{1}{\sqrt{N}}\beta \max_{i=1,\dots,2^N} X_i$$

if we set $\xi = \max_{i=1,\dots,2^N} X_i$ we get

$$\begin{aligned}\int_{|F_N|^P>\alpha} |F_N|^P d\mathbb{P} &= \int_{F_N>\alpha^{\frac{1}{p}}} |F_N|^P d\mathbb{P} + \int_{F_N\leq-\alpha^{\frac{1}{p}}} (-F_N)^P d\mathbb{P} \\ &\leq \int_{(\frac{\beta\xi}{\sqrt{N}}+\log 2)>\alpha^{\frac{1}{p}}} (\frac{\beta\xi}{\sqrt{N}} \log 2)^p d\mathbb{P} + \int_{\frac{\beta\xi}{\sqrt{N}}<-\alpha^{\frac{1}{p}}} (-\frac{\beta\xi}{\sqrt{N}})^p d\mathbb{P}\end{aligned} \tag{1.56}$$

Since $F_N > \alpha^{\frac{1}{p}} \Rightarrow \frac{\beta\xi}{\sqrt{N}}+\log 2 > \quad F_N > \alpha^{\frac{1}{p}}$ and $F_N \leq -\alpha^{\frac{1}{p}} \Rightarrow \frac{\beta\xi}{\sqrt{N}} \leq F_N \leq -\alpha^{\frac{1}{p}}$.

Now

$$\int_{(\frac{\beta\xi}{\sqrt{N}}+\log 2)>\alpha^{1/p}} (\frac{\beta\xi}{\sqrt{N}} + \log 2)^P d\mathbb{P} = \int_{\frac{\beta\xi}{\sqrt{N}}>\alpha^{1/p}-\log 2} (\frac{\beta\xi}{\sqrt{N}} + \log 2)^P d\mathbb{P} \tag{1.57}$$

assuming $\alpha^{\frac{1}{p}} - \log 2 > 0$.

$$\begin{aligned}&= \sum_{\ell=1}^{\infty} \int_{\ell(\alpha^{\frac{1}{p}}-\log 2)\leq\frac{\beta\xi}{\sqrt{N}}\leq(\alpha^{\frac{1}{p}}-\log 2)\ell+1} (\frac{\beta\xi}{\sqrt{N}} + \log 2)^P d\mathbb{P} \\ &\leq \sum_{\ell=1}^{\infty}((\ell+1)\alpha^{\frac{1}{p}} + \log 2)^P \mathbb{P}(\xi \geq \frac{\ell\sqrt{N}(\alpha^{\frac{1}{p}} - \log 2)}{\beta})\end{aligned} \tag{1.58}$$

On the other hand

$$\int_{\frac{\beta\xi}{\sqrt{N}}<-\alpha^{\frac{1}{P}}} (-\frac{\beta\xi}{\sqrt{N}})^P dIP = \sum_{\ell=1}^{\infty} \int_{(\ell+1)\alpha^{\frac{1}{P}} \leq \frac{\beta\xi}{\sqrt{N}} < -(\ell)\alpha^{\frac{1}{P}}} (-\frac{\beta\xi}{\sqrt{N}})^P dIP$$
$$\leq \sum_{\ell=1}^{\infty} (\ell+1)^P \alpha IP(\xi \leq -\frac{\alpha^{\frac{1}{P}} \ell\sqrt{N}}{\beta}) \tag{1.59}$$

Now on one hand

$$IP(\xi \geq \frac{\ell\sqrt{N}(\alpha^{\frac{1}{p}} - \ell n2)}{\beta}) \leq 2^N IP(X \geq \frac{\ell\sqrt{N}(\alpha^{\frac{1}{p}} - \ell n2)}{\beta})$$
$$\leq 2^N e^{-\frac{\ell^2 N}{2\beta^2}(\alpha^{\frac{1}{p}} - \ell n2)^2} \tag{1.60}$$

and on the other hand

$$IP(\xi \leq -\frac{\alpha^{\frac{1}{p}} \ell\sqrt{N}}{\beta}) \leq e^{-\frac{\ell^2 N}{2\beta^2}\alpha^{\frac{2}{p}}} \leq 2^N e^{-\frac{\ell^2 N}{2\beta^2}\alpha^{\frac{2}{p}}} \tag{1.61}$$

Let us estimate

$$\sum_{\ell=1}^{\infty}((\ell+1)\alpha^{\frac{1}{p}}) + \log 2)^p 2^N e^{-\frac{\ell^2_N}{2\beta_2}(\alpha^{\frac{1}{p}} - \ell n2)^2}$$
$$\leq 2^p(\log 2)^p \sum_{\ell=1}^{\infty} 2^N e^{-\frac{\ell^2 N}{2\beta^2}(\alpha^{1/p} - \ell n2)^2} + 2^p \alpha \sum_{\ell=1}^{\infty} (\ell+1)^p 2^N 2^{-\frac{\ell^2 N}{2\beta^2}(\alpha^{\frac{1}{p}} - \ell n2)^2} \tag{1.61}$$

if

$$\frac{(\alpha^{\frac{1}{p}} - \ell n2)^2}{2\beta^2} > 2\log 2$$

the first series in the r.h.s. of 1.62 does not exceed

$$\sum_{\ell=1}^{\infty} e^{-\frac{\ell^2 N(\alpha^{\frac{1}{p}} - \ell n2)^2}{4\beta^2}} \leq \sum_{\ell=1}^{\infty} e^{-\frac{\ell^2(\alpha^{\frac{1}{p}} - \ell n2)^2}{4\beta^2}} \tag{1.62}$$

which goes to zero when α goes to infinity.

The second series in the r.h.s. of 1.62 does not exceed

$$\alpha \sum_{\ell=1}^{\infty} (\ell+1)^P e^{-\ell^2(\frac{\alpha^{\frac{1}{p}} - \ell n2)^2}{4\beta^2}} \tag{1.63}$$

which goes to zero when α goes to infinity.

By similar argument it can be proved that

$$\lim_{\alpha\to\infty}\sup_{N\geq 1}\sum_{\ell=1}^{\infty}(\ell+1)^p\alpha I\!P(\xi\leq -\alpha^{\frac{1}{p}}\frac{\ell\sqrt{N}}{\beta})=0 \tag{1.64}$$

and this end the proof of the uniform $I\!L^P$ integrability which implies, taking into account of the convergence $I\!P$ almost everywhere, the convergence in all $I\!L^p(\Omega,\Sigma,I\!P)$.

Proof. of Theorem 1.1. The basic idea in the proof of the theorem 1.3 was that

$$\lim_{N\to\infty}\frac{1}{N}\log Z_N=\lim_{N\to\infty}\frac{1}{N}\log I\!E(Z_N1\!\!1_{\{\forall i\}}) \tag{1.65}$$

where

$$1\!\!1_{\{\forall i\}}=1\!\!1_{\{\forall i\in 1,\dots,2^N,|X_i|\leq\sqrt{2(1+\varepsilon)N\log 2}\}}$$

and $1\!\!1_{\{\forall i\}}=1$ almost everywhere.

(Since $1\!\!1_{\{\forall i\}}+1\!\!1_{\{\exists i\}}=1$ this implies $1\!\!1_{\{\exists i\}}=0$ almost everywhere).

The first problem is to find for the GREM the analogue of the interval $|X_i|\leq\sqrt{2(1+\varepsilon)N\log 2}$. The answer is the following. Let

$$A(1,N)=\{\omega/\forall j_1\in 1,\dots,\alpha_1^N,(\varepsilon^1_{j_1})^2\leq 2N\log\alpha_1+2\delta_1N\} \tag{1.66}$$

$$\begin{aligned}A(2,N)=&\{\omega/\forall j_1\in 1,\dots,\alpha_1^N,\forall j_2\in 1,\dots,\alpha_2^N,\\&(\varepsilon^1_{j_1})^2+(\varepsilon^2_{j_1,j_2})^2\leq 2N[\log\alpha_1+\log\alpha_2+\delta_2]\}\end{aligned} \tag{1.67}$$

and more generaly if $j\in 1,\dots,n$ and $\underline{\delta}=(\delta_1,\dots,\delta_n)$ is a positive vector

$$Q_j=\sum_{i=1}^{j}\log\alpha_i+\delta_j$$

$$A(j,N)=\{\omega\in\Omega/\forall k_1,\dots,\forall k_j;\quad\sum_{i=1}^{j}(\varepsilon^i_{k_1,\dots,k_j})^2\leq 2Q_jN\} \tag{1.68}$$

Let $X(Q_j,N)=1\!\!1_{\{A(j,N)\}}$ the characteristic function of the set $A(j,N)$ and $\prod_{j=1}^{n}X(Q_j,N)$ the characteristic function of the intersection of all sets $A(j,N)$ for $j=1,\dots,n$.

The first proposition is the analogue of the fact that

$$\max_{i=1,\ldots,2^N} |X_i| \leq \sqrt{2N(1+\varepsilon)\log 2}$$

Proposition. For any $\underline{\delta} = (\delta_1, ..., \delta_n) \in \mathbb{R}^n$, $\delta_i > 0$ $\forall i = 1, ..., n$ there exists $\Omega_1 \subset \Omega$ such that $\mathbb{P}(\Omega_1) = 1$ and for any Ω_1, $\exists N_1(\omega, \underline{\delta})$ such that, for any $N > N_1(\omega, \underline{\delta})$

$$\prod_{j=1}^{n} X(Q_j, N) = 1$$

Proof.

$$\begin{aligned} \mathbb{P}(\prod_{j=1}^{n} X(Q_j, N) = 0) &= \mathbb{P}(\bigcup_{j=1}^{n} A^c(j,N)) \\ &\leq \sum_{j=1}^{n} \mathbb{P}(A^c(j,N)) \end{aligned} \tag{1.69}$$

On the other hand,

$$\begin{aligned} \mathbb{P}(A^c(j,N)) &= \mathbb{P}(\max_{k_i,\ldots,k_j} \sum_{i=1}^{j} (\varepsilon^i_{k_1,\ldots,k_i})^2 \geq 2Q_j N) \\ &\leq e^{-\lambda Q_j N} \mathbb{E}(e^{\lambda \max_{k_1,\ldots,k_j} \sum_{i=1}^{j} \frac{1}{2}(\varepsilon^i_{k_1,\ldots,k_i})^2}) \\ &\leq e^{-\lambda Q_j N} \mathbb{E}(\max_{k_1,\ldots,k_j} e^{\lambda \sum_{k_1,\ldots,k_j}^{j} \frac{1}{2}(\varepsilon^i_{k_1,\ldots,k_i})^2}) \\ &\leq e^{-\lambda Q_j N} \alpha_1^N \cdots \alpha_j^N (\frac{1}{1-\lambda})^{\frac{\delta}{2}} \end{aligned} \tag{1.70}$$

Since

$$Q_j = \sum_{i=1}^{j} \ell_n \alpha_i + \delta_j$$

if we choose $\lambda = 1 - \eta$ we get that the r.h.s. of 1.71 is

$$(\eta)^{-\frac{j}{2}} \exp N[\eta \sum_{i=1}^{j} \log \alpha_i - (1-\eta)\delta_j] \tag{1.71}$$

since if

$$\eta < \frac{\delta_j}{2(\sum \log \alpha_i + \delta_j)}$$

$\Rightarrow \eta \Sigma \log \alpha_i - (1-\eta)\delta_j \leq -\frac{\delta_j}{2}$ choosing

$$\eta < \min_{1<j<n} \frac{\delta_j}{2(\delta_j + \sum_{i=1}^{j} \log \alpha_i)} \tag{1.72}$$

we get

$$\mathbb{P}(A^c(j,N)) \leq (\eta)^{\frac{j}{2}} e^{-\frac{\delta_j N}{2}} \tag{1.73}$$

Therefore (η, δ_j and n are fixed)

$$\sum_N \sum_{j=1}^{n} \mathbb{P}(A^c(j,N)) < +\infty \tag{1.74}$$

and we conclude by the Borel Cantelli lemma.

The second fact is a trivial application of the Markov inequality:

$$\exists \Omega_2 \subset \Omega \text{ such that } \mathbb{P}(\Omega_2) = 1, \text{ for any } \omega \in \Omega_2$$

and any $\gamma > 0$ $\exists N_2(\omega, \gamma)$ such that if $N > N_1(\omega, \gamma)$ then

$$Z_{n,N}(\beta) \leq \frac{e^{\gamma N} 2^N}{(\sqrt{2\pi})^n} \int_{\substack{\{\epsilon_1^2 + \cdots + \epsilon_n^2 \leq 2Q_n N\} \\ \{\epsilon_1^2 + \cdots + \epsilon_{n-1}^2 \leq 2Q_{n-1} N\} \\ \vdots \\ \{\epsilon_1^2 \leq 2Q_1 N\}}} d\varepsilon_1 \cdots d\varepsilon_n e^{-\sum_{i=1}^{n} (\frac{\epsilon_i^2}{2}) \beta \sum_{i=1}^{n} \sqrt{N q_i} \delta_i} \tag{1.75}$$

In fact by the Markov inequality

$$\mathbb{P}((\prod_{j=1}^{n} \chi(Q_1, N)) Z_{n,N}(\beta) \geq e^{\gamma N} \mathbb{E}(Z_{n,N}(\beta) \prod_{j=1}^{n} \chi(Q_j, N))) \leq e^{-\gamma N} \tag{1.76}$$

and the result follows from the Borel Cantelli Lemma.

Let us define:

$$\tilde{\xi}(k,\delta) = \{(x_i) \in I\!R^k | \forall j = 1,...,k; \sum_{i=1}^{j} \frac{(X_i)^2}{2} \leq Q_j\} \tag{1.77}$$

for $k = 1, 2, ..., n$

$\tilde{\xi}(n,\delta)$ is a compact subset of $I\!R$ as an intersection of a compact $S(n) = \{(X_i) \in I\!R^n / \sum_{i=1}^{n} \frac{(X_i)^2}{2} \leq Q_n\}$ and closed subsets $\bigcap_{j=1}^{n-1} S(j)$.

$\tilde{\xi}(n,\delta)$ is convex as an intersection of convex subset. Since $\tilde{\xi}(n,\delta)$ is compact, if $m^*(\beta) \equiv (\beta\sqrt{a_i})_{i=1}^n$ there exists an unique point $m(\delta) \in \tilde{\xi}(n,\delta)$ such that

$$\begin{aligned} \mathrm{dist}(m^*(\beta), \tilde{\xi}(n,\delta)) &= \inf_{\varepsilon \in \tilde{\xi}(n,\delta)} \mathrm{dist}(m^*(\beta), \varepsilon) \\ &= \mathrm{dist}(m^*(\beta), m(\delta)) \end{aligned} \tag{1.78}$$

where $\mathrm{dist}(x,y) = (\sum_{i=1}^{n} |x_i - y_i|^2)^{\frac{1}{2}}$.

Moreover, using the following identity

$$(\varepsilon - m^*) \cdot (\varepsilon - m^*) = ((\varepsilon - m) - (m^* - m)) \cdot ((\varepsilon - m) - (m^* - m)) \tag{1.79}$$

we get:

$$(\varepsilon - m^*)^2 = (\varepsilon - m)^2 + (m^* - m)^2 - 2(\varepsilon - m) - (m^* - m) \tag{1.80}$$

Now consider the linear form $\mathcal{L}(\varepsilon) = (\varepsilon - m) \cdot (m^* - m)$, $\mathcal{L}(\varepsilon) = 0$ is the equation of the hyperplane orthogonal to the vector $(m^* - m)_{i=1}^n$ passing by the point m. It is easy to check that since $\tilde{\xi}(n,\delta)$ is convex, if m^* is outside $\tilde{\xi}(n,\delta)$ then $\mathcal{L}(\varepsilon) \leq 0$, therefore

$$(\varepsilon - m^*)^2 \geq (\varepsilon - m)^2 + (m^* - m)^2 \tag{1.81}$$

if m^* is inside $\tilde{\xi}(n,\delta)$ then $m^* = m$ which implies

$$(\varepsilon - m^*)^2 = (\varepsilon - m)^2 = (\varepsilon - m)^2 + (m^* - m)^2 \tag{1.82}$$

We have to estimate

$$2^N e^{\gamma_N} (\frac{1}{\sqrt{2\pi}})^n \int_{\substack{\varepsilon_1 \leq 2Q_1 N \\ \vdots \\ \varepsilon_1^2 + \cdots + \varepsilon_n^2 \leq 2Q_n N}} d\varepsilon_1 \cdots d\varepsilon_n e^{-\frac{1}{2}\sum_{i=1}^{n} \varepsilon_i^2 + \beta \sum_{i=1}^{n} (Na_i)^{\frac{1}{2}} \varepsilon_i} \tag{1.83}$$

Since

$$\left.\begin{array}{c}\varepsilon_1^2 \le 2Q_1 N \\ \vdots \\ \varepsilon_1^2 + \cdots + \varepsilon_n^2 \le 2Q_n N\end{array}\right. \Leftrightarrow \quad \xi_{\tilde{\xi}(n,\delta)}(\tfrac{\varepsilon}{\sqrt{N}}) = 1$$

Setting $\varepsilon_i = \sqrt{N}\tilde{\varepsilon}_i$ we get that 1.84 can be written

$$2^N e^{\gamma N}(\frac{1}{\sqrt{N2\pi}})^n \int d\tilde{\varepsilon}_1 \cdots d\tilde{\varepsilon}_n \chi_{\tilde{\xi}(n,\delta)}(\tilde{\varepsilon}) e^{-\frac{1}{2}N\sum_{i=1}^n \tilde{\varepsilon}_i^2 + \beta N \sum_{i=1}^n (a_i)^{\frac{1}{2}}\tilde{\varepsilon}_i} \tag{1.84}$$

Now

$$\begin{aligned} &-\frac{1}{2}\sum_{i=1}^n \tilde{\varepsilon}_i + \beta \sum_{i=1}^n \sqrt{a_i}\tilde{\varepsilon}_i \\ &= -\frac{1}{2}(\tilde{\varepsilon} - m^*(\beta))^2 + \frac{1}{2}(m^*(\beta))^2 \\ &\le -\frac{1}{2}(\tilde{\varepsilon} - m)^2 - \frac{1}{2}(m^* - m)^2 + \frac{1}{2}(m^*(\beta))^2 \end{aligned} \tag{1.85}$$

where we have used 1.82 and 1.83, therefore 1.85 does not exceed.

$$\begin{aligned} &2^N e^{\gamma N}(\frac{1}{\sqrt{2\pi N}})^n e^{-\frac{1}{2}N(m^\alpha - m)^2 + \frac{1}{2}(m^\alpha)^2 N} \\ &\int d\tilde{\varepsilon}_i \cdots d\tilde{\varepsilon}_n X_{\tilde{\varepsilon}(n,\delta)}(\tilde{\varepsilon}) e^{-\frac{1}{2}N\sum_{i=1}^n (\varepsilon_i - m_i)^2} \\ &= 2^N e^{\gamma N} e^{\frac{1}{2}N[-\|m - m^*\|^2 + \|m^*\|^2]} \\ &(\frac{1}{\sqrt{2\pi}})^n \int d\varepsilon_1 \cdots d\varepsilon_n X_{\tilde{\varepsilon}(n,\delta)}(\frac{\varepsilon}{\sqrt{N}}) e^{-\frac{1}{2}\sum_{i=1}^n (\varepsilon_i - \sqrt{N}m_i)^2} \\ &\le 2^N e^{\gamma N} e^{\frac{1}{2}N(-\|m - m^*\|^2 + \|m^*\|^2)} \end{aligned} \tag{1.86}$$

Since the previous integral does not exceed $(\sqrt{2\pi})^n$.

Therefore we have proved the half of theorem 1.1.

Namely: for almost ω there exist $N(\omega)$ such that $N \ge N(\omega)$ implies:

$$F_{n,N}(\beta) \le \log 2 - \frac{1}{2}\|m - m^*\|^2 + \frac{1}{2}\|m^*\|^2 + \gamma \tag{1.87}$$

Now we want to prove a lower bound on the form:

$$F_{n,N}(\beta) \ge F(\beta) - \varepsilon$$

Inspirated by the case of the REM we minimize the partition function by taking only the contribution that comes from a neighbourhood of the point m, defined previously, namely we write

$$Z_{n,N}(\beta) \geq \exp\beta\sqrt{N} \inf_{\substack{\varepsilon^i \in \Delta i \\ i=1,\dots,n}} \sum_{i=1}^{n} \sqrt{a_i}\varepsilon_i$$
$$\sum_{k_1=1}^{\alpha_1^N} \mathbb{1}_{\Delta_1}(\varepsilon^1_{k_1}) \sum_{k_2=1}^{\alpha_2^N} \mathbb{1}_{\Delta_2}(\varepsilon^2_{k_1,k_2}) \cdots \sum_{k_n=1}^{\alpha_n^N} \mathbb{1}_{\Delta_n}(\varepsilon^2_{k_1,\dots,k_n}) \tag{1.88}$$

and choosing $\Delta_i = \{\sqrt{N}(m_i - 2\rho i) < \varepsilon_i < \sqrt{N}(m_i - \rho_i)\}$ we get

$$\inf_{\varepsilon^i \in \Delta_i} \sum_{i=1}^{n} \sqrt{a_i}\varepsilon_i = \sqrt{N}\sum_{i=1}^{n} \sqrt{a_i}m_i - 2\sqrt{N}\sum_{i=1}^{n} \rho_i m_i \tag{1.89}$$

The non trivial fact is to prove if $0 < \eta < 1$

$$\sum_{k_1=1}^{\alpha_1^N} \mathbb{1}_{\Delta_1}(\varepsilon^1_{K_1}) \sum_{K_n=1}^{\alpha_n^N} \mathbb{1}_{\Delta_n}(\varepsilon^n_{K_1,\dots,K_n}) \geq (1-\eta)\prod_{i=1}^{n} \alpha_i^N \mathbb{E}(\mathbb{1}_{\Delta i}) \tag{1.90}$$

almost everywhere.

In fact by using

$$\mathbb{P}(\sum \mathbb{1}_{\Delta_1} \dots \sum \mathbb{1}_{\Delta_n} \leq (1-\eta)\prod_{i=1}^{n} \alpha_1^n \mathbb{E}(\mathbb{1}_{\Delta_1}))$$
$$\leq \mathbb{P}(|\sum \mathbb{1}_{\Delta_1} \dots \sum \mathbb{1}_{\Delta_n} - \mathbb{E}(\sum \mathbb{1}_{\Delta_1} \dots \sum \mathbb{1}_{\Delta_n})| \geq \eta \prod_{i=1}^{n} \alpha_1^n \mathbb{E}(\mathbb{1}_1))$$
$$\leq \frac{1}{\eta^2[\prod_{i=1}^{n} i\alpha_i^N \mathbb{E}(\mathbb{1}\Delta_i)]^2} \mathbb{E}([\sum_{K_1} \mathbb{1}_{\Delta_1} \dots \sum_{K_n} \mathbb{1}_{\Delta_n} - \mathbb{E}(\sum \mathbb{1}_{\Delta_1} \dots \sum \mathbb{1}_{\Delta_n})]^2) \tag{1.91}$$

We have to take into account of cancellations that occur in

$$D_n^2 = \mathbb{E}([\sum \mathbb{1}_{\Delta_1} \dots \sum \mathbb{1}_{\Delta_n} - \mathbb{E}(\sum \mathbb{1}_{\Delta_1} \dots \sum \mathbb{1}_{\Delta_n})]^2) \tag{1.92}$$

For simplicity we will first consider the case $n = 2$

$$D^2 = \mathbb{E}(\sum_{k_1,k_2,\ell_1,\ell_2} [\mathbb{1}_{\Delta 1}(\varepsilon^1_{k_1})\mathbb{1}_{\Delta_2}(\varepsilon^2_{k_1,k_2}) - \mathbb{E}(\mathbb{1}_{\Delta_1}(\varepsilon^1_{k_1}))\mathbb{E}(\mathbb{1}_{\Delta_2}(\varepsilon^2_{k_1,k_2})]$$
$$\times [\mathbb{1}_{\Delta_1}(\varepsilon^1_{\ell_1})\mathbb{1}_{\Delta_2}(\varepsilon^2_{\ell_1,\ell_2}) - \mathbb{E}(\mathbb{1}_{\Delta_1}(\varepsilon^1_{\ell_1}))\mathbb{E}(\mathbb{1}_{\Delta_2}(\varepsilon^2_{\ell_1,\ell_2}))]$$
$$= \sum_{k_1,k_2,\ell_2,\ell_2} \mathbb{E}(\mathbb{1}_{\Delta_1}(\varepsilon^1_{k_1})\mathbb{1}_{\Delta_1}(\varepsilon^1_{\ell_1}))\mathbb{E}(\mathbb{1}_{\Delta_2})\varepsilon^2_{k_1,k_2})\mathbb{1}_{\Delta_2}(\varepsilon^2)) \tag{1.93}$$
$$- \mathbb{E}(\mathbb{1}_{\Delta_1}(\varepsilon^1_{k_1}))\mathbb{E}(\mathbb{1}_{\Delta_1}(\varepsilon^1_{\ell_1}))\mathbb{E}(\mathbb{1}_{\Delta_2}(\varepsilon^2_{k_1,k_2})\mathbb{E}(\mathbb{1}_{\Delta_2}(\varepsilon^2_{\ell_1,\ell_2}))$$

if $k_1 \neq \ell_1$ then $\varepsilon^1_{k_1}$, $\varepsilon^1_{\ell_2}$, $\varepsilon^2_{k_1,k_2}$, and $\varepsilon^2_{\ell_1,\ell_2}$ are independent and the corresponding terms are zero. Therefore

$$\sum_{k_1,k_2,\ell_1,\ell_2} \rightarrow \sum_{k_1,k_2,\ell_2}$$

writing

$$\sum_{k_1,k_2,\ell_2} = \sum_{k_1,k_2=\ell_2} + \sum_{k_1,k_2\neq\ell_2} \tag{1.94}$$

we get

$$\begin{aligned}
&\sum_{k_1,k_2} I\!E(1\!\!1_{\Delta_1}(\varepsilon^1_{k_1}))I\!E(1\!\!1_{\Delta_1}(\varepsilon^2_{k_1,k_2})) - [I\!E(1\!\!1_{\Delta_1}(\varepsilon^1_{k_1}))]^2[I\!E(1\!\!1_{\Delta_2}(\varepsilon^2_{k_1,k_2})]^2 \\
&= \sum_{K_1,K_2} I\!E(1\!\!1_{\Delta_1})I\!E(1\!\!1_{\Delta_2})[1 - I\!E(1\!\!1_{\Delta_1}I\!E(1\!\!1_{\Delta_2})] \\
&= \alpha_1^N\alpha_2^N I\!E(1\!\!1_{\Delta_1})I\!E(1\!\!1_{\Delta_2})[1 - I\!E(1\!\!1_{\Delta_2})I\!E(1\!\!1_{\Delta_2})]
\end{aligned} \tag{1.95}$$

for the first sum.

For the second sum we write

$$\begin{aligned}
&= \sum_{k_1}\sum_{k_2\neq\ell_2} I\!E(1\!\!1_{\Delta_1}(\varepsilon^1_{k_1}))I\!E(1\!\!1_{\Delta_2}(\varepsilon^2_{k_1,k_2}))I\!E(1\!\!1_{\Delta_2}(\varepsilon^2_{k_2,\ell_2})) \\
&- [I\!E(1\!\!1_{\Delta_1}(\varepsilon^1_{k_1}))]^2 I\!E(1\!\!1_{\Delta_2}(\varepsilon^2_{k_1,k_2}))I\!E(1\!\!1_{\Delta_2}(\varepsilon^2_{k_1,\ell_2})) \\
&= \sum_{k_1}\sum_{k_2\neq\ell_2} I\!E(1\!\!1_{\Delta_1}(\varepsilon^1_{k_1}))[1 - I\!E(1\!\!1_{\Delta_1})][I\!E(1\!\!1_{\Delta_2}]^2 \\
&= \alpha_1^N\alpha_2^N(\alpha_2^N - 1)I\!E(1\!\!1_{\Delta_1})[1 - I\!E(1\!\!1_{\Delta_1})[1 - I\!E(1\!\!1_{\Delta_1})](I\!E(1\!\!1_{\Delta_2}))^2
\end{aligned} \tag{1.96}$$

Let us now consider the general case

$$\begin{aligned}
D_n = &\sum_{\substack{k_1,\dots,k_n\\ \ell_1,\dots\ell_n}} I\!E(1\!\!1_{\Delta_1}(\varepsilon^1_{k_1})1\!\!1_{\Delta_1}(\varepsilon^1_{\ell_1}))I\!E(1\!\!1_{\Delta_2}(\varepsilon^2_{k_1 k_2})1\!\!1_{\Delta_2}(\varepsilon^2_{\ell_1,\ell_2}))\cdots \\
&I\!E(1\!\!1_{\Delta_n}(\varepsilon^n_{k_1,\dots,k_n})1\!\!1_{\Delta_n}(\varepsilon^n_{k_1,\dots,k_n})) - [I\!E(1\!\!1_{\Delta_1})]^2\cdots[I\!E(1\!\!1_{\Delta_n})]^2
\end{aligned} \tag{1.97}$$

If $k_1 \neq \ell_1$ the previous sums give zero, since in this case all the variables $\varepsilon^j_{k_1,\dots,k_j}$ and $\varepsilon^j_{\ell_1,\dots,\ell_j}$ (for any $j = 1,\dots,n$) are independent.

Writing

$$\sum_{k_1}\sum_{\substack{k_2,\dots,k_n\\ \ell_2,\dots,\ell_n}} = \sum_{k_1}\Big(\sum_{k_2\neq\ell_2}\sum_{\substack{k_3,\dots,k_n\\ \ell_3,\dots,\ell_n}} + \sum_{k_2=\ell_2}\sum_{\substack{k_3,\dots,k_n\\ \ell_3,\dots,\ell_n}}\Big) \tag{1.98}$$

the sum with $k_2 \neq \ell_2$ gives

$$\sum_{k_1} \sum_{k_2 \neq \ell_2} \sum_{\substack{k_3,\ldots,k_n \\ \ell_3,\ldots,\ell_n}} I\!E(\mathbb{1}_{\Delta_1})(1 - I\!E(\mathbb{1}_{\Delta_1}))[(I\!E(\mathbb{1}_{\Delta_2})]^2 \cdots [I\!E(\mathbb{1}_{\Delta_n})]^2 \tag{1.99}$$

the sum with $k_2 = \ell_2$ gives:

$$\begin{aligned} &\sum_{k_1} \sum_{k_2} \sum_{\substack{k_3,\ldots,k_n \\ \ell_3,\ldots,\ell_n}} I\!E(\mathbb{1}_{\Delta_1}) I\!E(\mathbb{1}_{\Delta_2}) I\!E(\mathbb{1}_{\Delta_3}\mathbb{1}_{\Delta_3}) \cdots I\!E(\mathbb{1}_{\Delta_n}\mathbb{1}_{\Delta_n}) \\ &- [I\!E(\mathbb{1}_{\Delta_1})]^2 \cdots [I\!E(\mathbb{1}_{\Delta_n})]^2 \end{aligned} \tag{1.100}$$

Writing

$$\sum_{\substack{k_3,\ldots,k_n \\ \ell_3,\ldots,\ell_n}} = \sum_{k_3 \neq \ell_3} \sum_{\substack{k_4,\ldots,\ell_n \\ \ell_4,\ldots,\ell_n}} + \sum_{k_3 = \ell_3} \sum_{\substack{k_4,\ldots,k_n \\ \ell_4,\ldots,\ell_n}}$$

the sum with $k_3 \neq \ell_3$ gives:

$$\begin{aligned} &\sum_{k_1} \sum_{k_2} (\sum_{k_3 \neq \ell_3} \sum_{\substack{k_4,\ldots,k_n \\ \ell_4,\ldots,\ell_n}} I\!E(\mathbb{1}_{\Delta_1}) I\!E(\mathbb{1}_{\Delta_2})(1 - I\!E(\mathbb{1}_{\Delta_1} I\!E(\mathbb{1}_{\Delta_2})) \\ &[I\!E(\mathbb{1}_{\Delta_1}]^2 \cdots [I\!E(\mathbb{1}_{\Delta_n})]^2 \end{aligned} \tag{1.101}$$

the sum with $k_3 = \ell_3$ gives:

$$\begin{aligned} &\sum_{k_1,k_2,k_3} (\sum_{\substack{k_4,\ldots k_n \\ \ell_4,\ldots,\ell_n}} I\!E(\mathbb{1}_{\Delta_1}) I\!E(\mathbb{1}_{\Delta_2}) I\!E_{\Delta_3}) I\!E(\mathbb{1}_{\Delta_4}\mathbb{1}_{\Delta_4}) \cdots I\!E(\mathbb{1}_{\Delta_n}\mathbb{1}_{\Delta_n}) \\ &- [I\!E(\mathbb{1}_{\Delta_1})]^2 \cdots [I\!E(\mathbb{1}_{\Delta_n})]^2 \end{aligned} \tag{1.102}$$

By iterating the previous process we get:

$$\begin{aligned} &\sum_{\substack{k_1,\ldots,k_n \\ \ell_2,\ldots,\ell_n}} I\!E(\mathbb{1}_{\Delta_1}(\varepsilon^1_{k_1} \mathbb{1}_{\Delta_1}(\varepsilon^1_k)) \cdots I\!E(\mathbb{1}_{\Delta_n}(\varepsilon^n_{k_1,\ldots,k_n}) \mathbb{1}_{\Delta_n}(\varepsilon^n_{\ell_n,\ldots,\ell_n}) \\ &- (I\!E(\mathbb{1}_{\Delta_1}))^2 \cdots (I\!E(\mathbb{1}_{\Delta_n}))^2 \\ &= \sum_{j=1}^{n} \sum_{k_1,\ldots,k_j} \sum_{k_{j+1} \neq \ell_{j+1}} \sum_{\substack{k_{j+2},\ldots,k_n \\ \ell_{j+2},\ldots,\ell_n}} I\!E(\mathbb{1}_{\Delta_1}) \cdots I\!E(\mathbb{1}_{\Delta_j})[1 - I\!E(\mathbb{1}_{\Delta_1}) \cdots I\!E(\mathbb{1}_{\Delta_j})] \\ &[I\!E(\mathbb{1}_{\Delta_j+1}) \cdots I\!E(\mathbb{1}_{\Delta_n})]^2 \end{aligned} \tag{1.103}$$

with the convention that undefined summations are not present.

Namely the two last summation $j = n-1$, $j = n$ correspond to the following sums:

$$\sum_{k_1,\dots,k_{n-1}} \sum_{k_n \neq \ell_n} I\!E(1\!\!1_{\Delta_1}) \cdots I\!E(1\!\!1_{\Delta_{n-2}})[1 - I\!E(1\!\!1_{\Delta_1}) \cdots I\!E(1\!\!1_{\Delta_{n-1}})][I\!E(1\!\!1_{\Delta_n})]^2$$

$$+ \sum_{k_1,\dots,k_n} I\!E(1\!\!1_{\Delta_1}) \cdots I\!E(1\!\!1_{\Delta_n})[1 - I\!E(1\!\!1_{\Delta_1}) \cdots I\!E(1\!\!1_{\Delta_n})]). \tag{1.104}$$

Using $1 - I\!E(1\!\!1_{s_1}) \cdots I\!E(1\!\!1(1\!\!1_{s_i}) \leq 1$ we get

$$D_n^2 \leq \sum_{j=1}^{n} (\alpha_1^N I\!E(1\!\!1_{\Delta_1})) \cdots (\alpha_j^N I\!E(1\!\!1_{s_j}))(\alpha_{j+1}^N I\!E(1\!\!1_{\Delta_1}))^2 \cdots (\alpha_n^N I\!E(1\!\!1_{\Delta_n}))^2 \tag{1.105}$$

where the term with $j = n$ is $\prod_{i=1}^{n} \alpha_i^N I\!E(1\!\!1_{s_i})$. Therefore

$$D_n^2 \leq [\prod_{j=1}^{n} \alpha_j^N I\!E(1\!\!1_{\Delta_j})] \sum_{j=1}^{n} (\alpha_{j+1}^N I\!E(1\!\!1_{\Delta_{j+1}})) \cdots (\alpha_n^N I\!E(1\!\!1_{\Delta_n})) \tag{1.106}$$

with the convention that the term with $j = n$ is equal 1.

Now

$$\frac{D_n^2}{[\prod_{i=1}^{n} \alpha_1^N I\!E(1\!\!1_{\Delta_i})]^2} \leq \sum_{j=1}^{n} \frac{1}{\prod_{i=1}^{j} \alpha_i^N I\!E(1\!\!1_{\Delta_i})} \tag{1.107}$$

Therefore if $\prod_{i=1}^{j} \alpha_i^N I\!E(1\!\!1_{\Delta_i}) \geq \mathcal{F}_j(N)$ and

$$\sum_{N \geq 1} \sum_{j=1}^{n} [\mathcal{F}_j(N)]^{-1} < +\infty \tag{1.108}$$

Using the Borel Cantelli lemma we get that 1.50 is true.

If $\Delta_i = \{\sqrt{N}(m_i - 2\rho_i) < \varepsilon_i < \sqrt{N}(m_i - \rho_i)\}$

Then using

$$\int_x^y e^{-\frac{t^2}{2}} \frac{dt}{\sqrt{2\pi}} = \int_x^y \frac{t}{t} e^{-\frac{t^2}{2}} \frac{dt}{\sqrt{2\pi}} \tag{1.109}$$

Setting $\mu = \frac{1}{t}$ $v = -e^{-t^2/2}$ and performing an integration by part we get

$$\int_x^y e^{-\frac{t^2}{2}} \frac{dt}{\sqrt{2\pi}} = \frac{e^{-\frac{x^2}{2}}}{x\sqrt{2\pi}} - \frac{e^{-\frac{y^2}{2}}}{y\sqrt{2\pi}} - \int_x^y \frac{e^{-\frac{t^2}{2}}}{t^2} \frac{dt}{\sqrt{2\pi}} \tag{1.110}$$

Since

$$\int_x^y \frac{e^{-\frac{t^2}{2}}}{t^2} \frac{dt}{\sqrt{2\pi}} \leq \frac{1}{x^2\sqrt{2\pi}} \int_x^\infty e^{-\frac{t^2}{2}} dt$$

$$\leq \frac{1}{x^3\sqrt{2\pi}} e^{-x^2/2} \tag{1.111}$$

We get

$$\int_x^y e^{-t^2/2} \frac{dt}{\sqrt{2\pi}} \geq \frac{e^{-x^2/2}}{x\sqrt{2\pi}} [1 - \frac{1}{x^2} - \frac{x}{y} e^{-\frac{1}{2}(x-y)(y+x)}] \tag{1.113}$$

in our case

$$\frac{x}{y} = \frac{m_i - \rho_i}{m_i - 2\rho_i}$$

and $(x-y)(x+y) = -N\rho_i(2m_i - 3\rho_i)$ therefore if N is large enough

$$I\!E(1\!\!1_{\Delta_i}(\varepsilon_i)) \geq \frac{e^{-\frac{1}{2}N(m_i - 2\rho_i)^2}}{2\sqrt{N}(m_i - \rho_i)\sqrt{2\pi}} \tag{1.114}$$

Therefore

$$\prod_{i=1}^{j} \alpha_i^N I\!E(1\!\!1_{\Delta_i}) \geq (\frac{1}{2(2\pi N)^{1/2}})^j \frac{\exp[\sum_{i=1}^{j} 2N(\rho_i m_i - \rho_i^2)]}{\prod_{i=1}^{j}(m_i - \rho_i)} \tag{1.115}$$

where we have used

$$(\prod_{i=1}^{j} \alpha_i^N) e^{-\sum_{i=1}^{j} \frac{N}{2} m_i^2} \geq 1$$

Since $(m_i)_{i=1}^n \in \xi(n) \Rightarrow \forall j \quad \sum_{i=1}^{j} \frac{m_i^2}{2} \leq \sum_{i=1}^{j} \log \alpha_i$.

If n is fixed, ρ_i small enough (in order $\rho_i m_i - \rho_i^2 > 0$) we get

$$\sum_N \sum_{j=1}^n (\mathcal{F}_j(N))^{-1} < +\infty \tag{1.116}$$

Now using 1.91 y 1.90 and 1.116 we get for almost all ω there exists $N(\omega)$ such that for any $N \geq N(m)$

$$Z_{n,N}(\beta) \geq (1-\eta)\exp[N\{\sum_{i=1}^n [\log \alpha_i - \frac{1}{2}(m_i - 2\rho_i)^2 + \beta\sqrt{a_i}(m_i - 2\rho_i)]\}]$$
$$= (1-\eta)\exp[N[\log 2 + \sum_{i=1}^n \beta\sqrt{a_i}m_i - \frac{1}{2}(m_i)^2]] \cdot \exp[2N\sum_{i=1}^n (\rho_i m_i - \rho_i^2 - \beta\sqrt{a_i}\rho_i)] \tag{1.117}$$

Now since $m^* \equiv (\beta\sqrt{a_i})_{i=1}^n$ we get

$$\sum_{i=1}^n \beta\sqrt{a_i}m_i - \frac{1}{2}(m_i)^2 = < m^*, m > -\frac{1}{2}\|m\|^2$$
$$= \frac{1}{2}(\|m^*\|^2 - \|m - m^*\|^2)$$

therefore

$$Z_{n,N}(\beta) \geq (1-\eta)\exp NF_n(\beta)\exp 2N\sum_{i=1}^n (\rho_i m_i - \rho_i^2 - \beta\sqrt{a_i}\rho_i) \tag{1.118}$$

choosing

$$-2\sum_{i=1}^n \rho_i m_i + \rho_i^2 + \beta\sqrt{a_i}\rho_i \leq \varepsilon$$

we get the result.

The proof of the uniform $I\!\!L^p$ integrability is very similar to the case of the REM and is left to the reader.

2. Explicit Evaluation of the Free Energy the GREM

This paragraph is also a correction of a similar one in [8] where the number of misprints make the reading involved.

Let $(a_i)_{i=1}^n$, $(\log \alpha_i)_{i=1}^n$ two sequences of real positive numbers such that $\sum_{i=1}^n \log \alpha_i = \log 2$ and $\sum_{i=1}^n a_i = 1$.

Let

$$\tilde{\xi}(n) = \{\varepsilon \in I\!R^n / \forall j \in 1, 2, ..., n \quad \sum_{i=1}^{j} \frac{\varepsilon_i^2}{2} \leq \sum_{i=1}^{j} \log \alpha_i\}$$

Let $m^* \equiv (\beta\sqrt{a_i})_{i=1}^n$, we want to find explicitly the coordinates of the point m of $\tilde{\xi}(n)$ which is at minimal distance to m^*.

Let us define, if j and k are integers $1 \leq j \leq k \leq n$

$$B_{j,k} = \frac{2 \sum_{i=1}^{k} \log \alpha_i}{\sum_{i=j}^{k} a_i}$$

In order to avoid ambiguities we assume that all numbers $B_{j,k}$ are distints, by continuity we can always change the parameters α_i, a_i in such a way that this hypothesis is true and by a limiting procedure we will obtain the coordinate of m is the general case.

Let us define the following family of integers

$$\begin{aligned} J_0^* &= 0 \\ J_1^* &= \inf(1 \leq J \leq n-1 | B_{1,J} < B_{J+1,\ell} \quad \forall \ell \geq J+1) \\ &\quad \text{if such an integer exists} \\ &= n \text{ otherwise.} \end{aligned}$$

If $J_1^* < n$ we define.

$$\begin{aligned} J_2^* &= \inf(J_1^* < J \leq n-1 | B_{J_1^*+1,J} < B_{J+1,\ell} \quad \forall \ell \geq J+1) \\ &\quad \text{if such an integer exists} \\ &= n \text{ otherwise.} \end{aligned}$$

More generaly if $J_{k-1}^* < n$ we define

$$J_k^* = \inf(J_{k-1}^* < J \leq n-1 | B_{J_{k-1}^*+1,J} < B_{k+1,\ell} \forall \ell \geq J \geq 1)$$

if such an integer exists

$$J_k^* = n \text{ otherwise}$$

in which case we end the process of definition.

Let us define also $\ell(n)$ as the only integer $\ell(n) \in 1, 2, ..., n$ such that $J_{\ell(n)}^* = n$.

We define also for $k = 1, 2, ..., \ell(n)$

$$\beta_k^2 = B_{J_{k-1}^*+1, J_k^*}, \ \beta_0 = 0$$

and we remark that by definition of J_k^* the β_k are strictly increasing with k.

We will prove that if $\beta_k \leq \beta \leq \beta_{k+1}$ for $k = 0, 1, ..., \ell(n)$ then the point $\tilde{m}$ of $\mathbb{R}^n$ with coordinates

$$\tilde{m}_i = \beta_\ell \sqrt{a_i} \text{ if } i \in [J_{\ell-1}^* + 1, ..., J_\ell^*] \quad \text{for } \ell = 1, ..., k$$
$$\tilde{m}_i = \beta \sqrt{a_i} \text{ if } i \in [J_k^* + 1, ..., n]$$

is the point of the compact convex subset $\tilde{\xi}(n)$ at minimal distance from $m^* = (\beta\sqrt{a_i})_{i=1}^n$

We will first prove that $\tilde{m} \in \tilde{\xi}(n)$.

We will use the two following facts

Let $(x_i)_{i=1}^n$ a bounded family of real number $(y_i)_{i=1}^n$ a bounded family of strictly positive numbers.

Let

$$S_{j,k} = \frac{\sum_{i=j}^{k} x_i}{\sum_{i=j}^{k} y_i} \quad \text{for } 1 \leq j \leq k \leq m$$

Fact 1.

$$\min_{i=j,...,k} \left(\frac{x_i}{y_i}\right) \leq S_{j,k} \leq \max_{i=j,...,k} \left(\frac{x_i}{y_i}\right)$$

This follows from the identity

$$S_{j,k} = \frac{\sum_{i=1}^{k} y_i \left(\frac{x_i}{y_i}\right)}{\sum_{i=j}^{k} y_i}$$

Fact 2. Assuming all the $S_{j,k}$ distinct if

$$\inf_{1\le \ell\le n}(S_{1,\ell}) = S_{1,\ell^*}$$

then

$$\ell^* = \inf(1 \le \ell \le n-1 / S_{1,\ell} < S_{\ell+1,k}\ \forall k \ge \ell+1)$$

if such an integer exists

$$\ell^* = n \text{ otherwise.}$$

Proof. Let us call $\tilde{\ell}^*$ the only integer such that

$$S_{1,\tilde{\ell}^*} = \inf_{1\le \ell\le n}(S_{1,\ell}).$$

In particular $\forall \ell \ge \tilde{\ell}^*_{+1}$ we have $S_{1,\tilde{\ell}^*} < S_{1,\ell}$.
Using fact 1 we get

$$\min(S_{1,\tilde{\ell}^*}; S_{\tilde{\ell}^*+1,\ell}) \le S_{1,\ell} \le \max(S_{1,\tilde{\ell}^*}; S_{\tilde{\ell}^*+1,\ell})$$

therefore $S_{1,\tilde{\ell}^*} < S_{\tilde{\ell}^*_1,\ell}$ $\forall \ell \ge \tilde{\ell}^* + 1$, that is $\tilde{\ell}^1$ satisfies the inequality into the bracket of the definition of ℓ^*, ℓ^* being the infimum we get $\ell^* \le \tilde{\ell}^*$.
Let us assume $\ell^* < \tilde{\ell}^*$, since

$$S_{1,\tilde{\ell}^*} \ge \min(S_{1,\ell^*}; S_{\ell^*+1,\tilde{\ell}^*})$$

and by definition of ℓ^* $S_{1,\ell^*} < S_{\ell^*+1,\tilde{\ell}^*}$ we get $S_{1,\ell^*} \le S_{1,\tilde{\ell}^*}$; since $\ell^* < \tilde{\ell}^*$ and all $S_{j,k}$ are distincts this implies $S_{1,\ell^*} < S_{1,\tilde{\ell}^*}$ which contradicts the definition of $\tilde{\ell}^*$ therefore $\tilde{\ell}^* \le \ell^*$ and we get $\tilde{\ell}^* = \ell^*$ ■

Using the fact 2 it can be easily checked that

$$B_{1,J_1^*} < B_{1,k} \quad \forall k \in 1, ..., n$$

$$B_{J_1^*+1,J_2^*} < B_{J_1^*+1,k} \quad \forall k \in J_1^*+1, ..., n$$

and more generaly for $k \in 1, ..., \ell(n)$

$$B_{J_{k-1}^*+1,J_K^*} < B_{J_{k-1}^1+1,\ell} \quad \forall \ell \in J_{k-1+1,...,n}^*.$$

Let us now prove that for any $p \in 1, 2, ..., n$.

$$(\tilde{m}_i)_{i=1}^p \in \{\sum_{i=1}^p \frac{X_i^2}{2} \leq \sum_{i=1}^p \log \alpha_i\}$$

For a given $p \in 1, 2, ..., n$ there exists an unique integer $\hat{p} \in 1, ..., \ell(n)$ such that $p \in [J_{\hat{p}-1}^* + 1, ..., J_{\hat{p}}^*]$. We assume $\beta_K \leq \beta \leq \beta_{K+1}$ and consider the case $\hat{p} \leq k$. We have to prove that

$$\forall \ell \in 1, ..., J_1^* \qquad \sum_{i=1}^{\ell} \beta_1^2 a_i \leq 2 \sum_{i=1}^{\ell} \log \alpha_i$$

$$\forall k \in 2, ..., \hat{p} - 2$$
$$\forall \ell \in J_k^* + 1, ..., J_{k+1}^*$$
$$\sum_{i=1}^{J_1^*} \beta_1^2 a_i + \sum_{i=J_1^*+1}^{J_2^*} \beta_2^2 a_i + \cdots + \sum_{i=J_{K-1}^*+1}^{J_K^*} \beta_{K-1}^2 a_i + \sum_{i=J_K^*+1}^{\ell} \beta_K^2 a_i$$
$$\leq 2 \sum_{i=1}^{\ell} \log \alpha_1$$

and

$$\sum_{K=0}^{\hat{p}-2} \sum_{i=J_K^*+1}^{J_{K+1}^*} \beta_k^2 a_i + \sum_{i=J_{\hat{p}-1}+1}^{\ell} \beta_{\hat{p}}^2 a_i \leq 2 \sum_{i=1}^{\ell} \log \alpha_i$$

for $\ell \in J_{\hat{p}-1}^* + 1, ..., p$

The first inequality is true since

$$\beta_1^2 = B_{1,J_1^*} \leq B_{1,\ell} \quad \forall 1 \leq \ell \leq J_1^*$$

Since $\sum_{i=1}^{J_1^*} \beta_1^2 a_i = 2 \sum_{i=1}^{J_1^*} \log \alpha_i$, the inequality corresponding to $k = 2$ is equivalent to

$$\sum_{i=J_1^*+1}^{\ell} \beta_2^2 a_i \leq 2 \sum_{i=J_1^*+1}^{\ell} \log \alpha_i$$

which is true, since $\beta_2^2 = B_{J_1^*+1, J_2^*} < B_{J_1^*+1, \ell}$ $\forall \ell \in J_1^* + 1, J_2^*$ All the other inequality are proved in similar way. The case $\hat{p} > k$ is proved similarly.

Now starting from the identity

$$(\varepsilon - m^*)^2 - (\varepsilon - \tilde{m})^2 - (\tilde{m} - m^*)^2 = -2(\varepsilon - \tilde{m}) \cdot (m^* - m^2)$$

which follows by expanding the right hand side of the following identity

$$(\varepsilon - m^*)^2 \cdot (\varepsilon - m^*) = ((\varepsilon - \tilde{m}) - (m^* - \tilde{m})) \cdot ((\varepsilon - \tilde{m}) - (m^* - \tilde{m}))$$

If we can prove that for any $\varepsilon \in \tilde{\xi}(m)$

$$\mathcal{L}(\varepsilon) = (\varepsilon - \tilde{m}) \cdot (m^* - \tilde{m}) \leq 0$$

then $\tilde{m} = m$.

In fact $\mathcal{L}(\varepsilon) = 0$ is the equation of the hyperplane of codimension 1 orthogonal to the vector $(m^* - \tilde{m})$ passing by $\tilde{m}$, $\mathcal{L}(\varepsilon) \leq 0$ for all $\varepsilon \in \xi(m)$, means that all $\varepsilon \in \xi(m)$ belong to the half space defined by $\mathcal{L}(\varepsilon) \leq 0$, since $\tilde{m} \in \xi(n)$ this implies that $\tilde{m}$ is the point at minimal distance from m^*.

Now since $\tilde{m}_i = m_i^* \; \forall i \geq J_k^* + 1$

$$\begin{aligned}\mathcal{L}(\varepsilon) &= \sum_{i=1}^{J_k^*} (\varepsilon_i - \tilde{m}_i)(m_i^* - \tilde{m}_i) \\ &= \sum_{\ell=1}^{k} (\frac{\beta}{\beta\ell} - 1) \sum_{i=J_{\ell-1}^*+1}^{J_k^*} (\varepsilon_i - \tilde{m}_i)\tilde{m}_i\end{aligned}$$

Let us prove that if $\varepsilon \in \bigcap_{j=1}^{J_k^*} (S(j)) = \tilde{\xi}(J_k^*)$ then

$$\varepsilon \in \bigcap_{\ell=1}^{k} \{\varepsilon \in \mathbb{R}^n / \mathcal{G}_{J_\ell^*}(\varepsilon) \equiv \sum_{i=1}^{J_\ell^*} (\varepsilon_i - \tilde{m}_i)\tilde{m}_i \leq 0\}$$

It is sufficient to prove that for any $\ell \in 1, 2, \ldots, k$ if

$$\varepsilon \in S(J_\ell^*) \text{ then } \sum_{i=1}^{J_\ell^*} (\varepsilon_i - \tilde{m}_i)\tilde{m}_i \leq 0$$

Let us compute the maximum of $\sum_{i=1}^{J_\ell^*} (\varepsilon_i - \tilde{m}_i)\tilde{m}_i$ with the constraint

$$\sum_{i=1}^{J_\ell^*} \varepsilon_i^2 \leq 2 \sum_{i=1}^{J_\ell^*} \log \alpha_i$$

It is easy to check that the max is attained for a point which belongs to the sphere

$$\sum_{i=1}^{J_\ell^*} \varepsilon_i^2 = 2\sum_{i=1}^{J_\ell^*} \log \alpha_i$$

Let

$$L(\lambda) = \sum_{i=1}^{J_\ell^*}(\varepsilon_i - \tilde{m}_i)\tilde{m}_i + \lambda[\sum_{i=1}^{J_\ell^*} \frac{\varepsilon_i^2}{2} - \sum_{i=1}^{J_\ell^*} \log \alpha_i]$$

then

$$\frac{\partial L(\lambda)}{\partial \varepsilon_i} = \tilde{m}_i + \lambda \varepsilon_i$$

therefore

$$\frac{\partial L(\lambda)}{\partial \varepsilon_i} = 0$$

is equivalent to $\varepsilon_i^* = -\frac{\tilde{m}_i}{\lambda}$

Choosing λ such that

$$\sum_{i=1}^{J_\ell^*} \frac{\varepsilon_i^*}{2} = \sum_{i=1}^{J_\ell^*} \log \alpha_i$$

$$\text{i.e. } \frac{1}{\lambda^2}(\sum_{k=1}^{\ell} \sum_{i=J_{k-1}^*+1}^{J_k^*} \beta_k^2 a_i) = \sum_{i=1}^{J_\ell^*} \log \alpha_i$$

by definition of $\beta_1, ..., \beta_k$ we get $\lambda^2 = 1$.

Since

$$\frac{\partial^2 L}{\partial \varepsilon^2 i} = \lambda \quad \frac{\partial^2 L}{\partial \varepsilon_i \partial \varepsilon_j} = 0$$

$\lambda = -1$ is a maximum $\lambda = +1$ is a minimum.

Therefore $L(\lambda) \leq 0$ that is $\varepsilon \in S(J_\ell^*) \Rightarrow \mathcal{G}_{J_\ell^*}(\varepsilon) \leq 0$.

Now

$$\sum_{i=J_{\ell-1}^*+1}^{J_\ell^*} (\varepsilon_i - \tilde{m}_i)\tilde{m}_i = \mathcal{G}_{J_\ell^*} - \mathcal{G}_{J_{\ell-1}^*}$$

with the convention that $\mathcal{G}_{j_0^*} = 0$

Therefore

$$\begin{aligned}
\mathcal{L}(\varepsilon) &= \sum_{\ell=1}^{k}\left(\frac{\beta}{\beta\ell}-1\right)(\mathcal{G}_{J_\ell^*}-\mathcal{G}_{J_{\ell-1}^*}) \\
&= -\sum_{\ell=1}^{k}(\mathcal{G}_{J_\ell^*}-\mathcal{G}_{J_{\ell-1}^*})+\beta\sum_{\ell=1}^{k}\frac{1}{\beta_\ell}(\mathcal{G}_{J_\ell^*}-\mathcal{G}_{J_{\ell-1}^*}) \\
&= -\mathcal{G}_{J_k^*}+\beta\sum_{\ell=1}^{k-1}\left(\frac{1}{\beta_\ell}-\frac{1}{\beta_{\ell+1}}\right)\mathcal{G}_{J_\ell^*}+\frac{\beta}{\beta_k}\mathcal{G}_{J_k^*} \\
&= \left(\frac{\beta}{\beta_k}-1\right)\mathcal{G}_{J_k^*}+\beta\sum_{\ell=1}^{k-1}\left(\frac{1}{\beta_\ell}-\frac{1}{\beta_{\ell+1}}\right)\mathcal{G}_{J_\ell^*}
\end{aligned}$$

Since $\beta > \beta_k \Rightarrow \frac{\beta}{\beta_k}-1 \geq 0$ $\beta_\ell < \beta_{\ell+1} \Rightarrow \frac{1}{\beta_\ell}-\frac{1}{\beta_{\ell+1}} \geq 0$ therefore since $\varepsilon \in \tilde{\xi}(J_k^*) \Rightarrow \varepsilon \in \bigcap_{\ell=1}^{k}\{\varepsilon/\mathcal{G}_{j_\ell^*}(\varepsilon) \leq 0\}$ we get $\varepsilon \in \tilde{\xi}(J_k^*) \Rightarrow \mathcal{L}(\varepsilon) \leq 0$ ■

Therefore if $F_n(\beta) = \frac{1}{2}\|m^*\| - \frac{1}{2}\|m-m^*\|^2 + \log 2$ we get

$$\begin{aligned}
F_n(\beta) &= \frac{\beta^2}{2}+\log 2 \quad \text{if} \quad \beta \leq \beta_1 \\
&= \sum_{\ell=1}^{k}\beta_\ell\beta\Big(\sum_{i=J_{\ell-1}^*+1}^{J_\ell^*}a_i\Big)+\sum_{i=J_\ell^*+1}^{n}\Big(\frac{1}{2}\beta^2a_i+\log\alpha_i\Big) \quad \text{if} \quad \beta_k \leq \beta \leq \beta_{k+1} \\
&\qquad \forall k \in 1,2,\ldots,\ell(n) \\
&= \sum_{\ell=1}^{\ell(n)}\beta_\ell\Big(\sum_{i=J_{\ell-1}^*}^{J_\ell^*}a_i\Big) \quad \text{if} \quad \beta \geq \beta_{\ell(n)}
\end{aligned}$$

References

[1] Derrida, B., Random Energy Model: Limit of a Family of Disordered Model, *Phys. Rev. Lett.* **65**, 79-82 (1980).

[2] Derrida, B., Random Energy Model: An Exactly solvable Model of Disordered Systems, *Phys. Rev. B* **26**, 2613-2626 (1981).

[3] Derrida, B., A Generalization of the Random Energy Model which Includes Correlations between Energies, *J. Phys. Lett.* **46**, 2401-2407 (1985).

[4] Gardner, E., Solution of the Generalized Random Energy Model, *J. Phys. C* **15**, 5783 (1986).
[5] Kirkpatrick, S., D. Sherrington, Solvable Model of the Spin Glass, *Phys. Rev. Lett.* **35**, 1792-1796 (1975).
[6] Mezard, M., G. Parisi, M.A. Virasoro, Spin Glass Theory and Beyond, *World Scientific* (1980).
[7] Olivieri, E., P. Picco, On the Existence of Thermodynamics for the Random Energy Model, *Com. Math. Phys.* **96**, 125-144 (1984).
[8] Capocaccia, D., M. Cassandro, P. Picco, On the Existence of Thermodynamics for the Generalized Random Energy Model, *J. Stat. Phys.* **66**, 493-505 (1987).
[9] Galvez, A., S. Martínez, P. Picco, Fluctuation a Derrida's REM and GREM, *J. Stat. Phys.* (1989).
[10] Ruelle, D., A Mathematical Reformulation of Derrida's REM and GREM, *Com. Math. Phys.* **198**, 225-239 (1987).